IMAGES
of Aviation

MALMSTROM AIR FORCE BASE

On the Cover: Pictured is a KC-97 tanker, the *City of Great Falls*, and its flight crew on the ramp at Great Falls Air Force Base. (Courtesy 341st Missile Wing.)

IMAGES
of Aviation

MALMSTROM AIR FORCE BASE

Lt. Col. George A. Larson, USAF (Ret.)

ARCADIA
PUBLISHING

ISBN 978-1-4671-0548-4

Published by Arcadia Publishing
Charleston, South Carolina

Printed in the United States of America

Library of Congress Control Number: 2020933060

For all general information, please contact Arcadia Publishing:
Telephone 843-853-2070
Fax 843-853-0044
E-mail sales@arcadiapublishing.com
For customer service and orders:
Toll-Free 1-888-313-2665

Visit us on the Internet at www.arcadiapublishing.com

This book is dedicated to the men and women who served at Great Falls Army Air Base during World War II up to the current Malmstrom Air Force Base, with its intercontinental ballistic missiles, who have defended the freedom of the nation.

Contents

Acknowledgments

I want to thank the men and women of the 341st Missile Wing who assisted in my visit to and research on Malmstrom Air Force Base, especially Col. Ronald Allen, wing commander. The wing public affairs staff and historian Bill Medema provided photographs and historical information. The Malmstrom Air Force Base Museum provided World War II air defense and Cold War–era photographs. The 40th Helicopter Squadron flew me to the Lima-01 Launch Facility for onsite research and a tour, with personnel explaining their nuclear deterrence mission. Back on the base, the 341st Security Forces Tactical Response Team provided a demonstration of securing a launch facility during an unauthorized entry. The Great Falls Museum and 120th Montana Air National Guard Airlift Wing provided historical photographs. The US Air Force Historical Research Agency filled in information on the base up to 1962 and the activation of the Minuteman I intercontinental ballistic missile. Most importantly, the staff of the Air Force Book Program in New York authorized a research visit and the cooperation of Malmstrom Air Force Base. This book would not have been possible without all these dedicated professionals working to honor the men and women who served on this installation.

Introduction

Malmstrom Air Force Base (AFB) traces its beginnings to September 1, 1939, when German military units invaded western Poland, starting World War II in Europe, which eventually became a two-ocean war for the United States. With the outbreak of war in Europe, members of the Great Falls Chamber of Commerce contacted their state's US senators—Burton K. Wheeler and James E. Murray—requesting they lobby the US War Department for consideration of military installations in Montana. A US Army inspection group headed by Colonel Deford with Maj. Ralph Royce arrived at Great Falls to determine the feasibility of a military base. The chamber of commerce briefed them on the advantages the city could offer an airbase. It was pointed out that the location was ideal for a bomber training installation and for the Lend-Lease program, which supplied US allies with aircraft and supplies. In 1941, the group revisited Great Falls and surveyed the city and surrounding areas for possible airbase sites, although on December 20, 1941, chamber of commerce officials found out that the Great Falls base had been moved to the Army's "Dead File." Gen. Hap Arnold later ordered that plans for an airbase at Great Falls be considered again. General Arnold was aware that to fight a global war, the Army Air Forces were going to have to train hundreds of thousands of personnel.

The first indication that an airbase was coming to Great Falls was on April 12, 1942, when Bill Killgreen of Ellerbe & Company, consulting engineer, arrived and indicated that work would start that summer on an airbase near the city. On May 6, Senators Wheeler and Murray along with Rep. James F. O'Connor announced that the War Department had authorized construction of an Army Air Forces installation in Montana at a cost in excess of $3 million.

Construction started on June 8, 1942. The 2nd Air Force had responsibility for airbases in Montana: Great Falls, Cutbank, Glasgow, and Lewistown. The base east of Great Falls was referred to as "East Base" since the 7th Ferrying Group operated at Great Falls's municipal airport on Gore Hill, west of the city, on a high bluff overlooking the city. The 7th Ferry Group's mission was to establish an air route from Great Falls to Ladd Field at Fairbanks, Alaska, as part of the Lend-Lease program.

On July 3, 1942, Lt. Col. Robert W. Owen, former executive officer of Geiger Field at Spokane, Washington, arrived to assume command of the Great Falls Army Air Base for the 2nd Air Force, which was in charge of all heavy bombardment training for the Army. On October 21, the official opening of three satellite airbases was announced by Major Eaton. These airbases at Cutbank, Glasgow, and Lewistown were to be used for Boeing B-17 Flying Fortress training. The first B-17 landed on Great Falls Army Air Base's snow-covered runway on November 30, 1942. The base served as headquarters for four bomb groups from November 1942 to October 1943, and was responsible for one of the bomb group's four bomb squadrons. Upon completion of B-17 bomb group training in October 1943, Great Falls Army Air Base was transferred to Air Transport Command, and units from Gore Field transferred to the base.

The Lend-Lease program continued, and included the processing of Bell P-39 Airacobras, Douglas C-47 Skytrains, North American B-25 Mitchells, and Douglas A-20 Havocs. B-25s arrived by rail and were assembled on the base, with other aircraft flown in by military and Women's Air Force Service Pilots (WASPs). These aircraft were flown by Army pilots from Great Falls on the Alaskan-Siberian route through Canada to Fairbanks, Alaska, and transferred to waiting Soviet air force pilots, who flew them to bases in Siberia, with approximately 8,000 aircraft processed and shipped during a 21-month period.

In September 1947, the US Air Force became an independent military service, and the base's name changed to Great Falls Air Force Base. The Cold War heated up after the Soviet Union closed land travel routes between East Germany and the Allied-controlled sectors of Berlin. On June 25, 1948, Operation Vittles, the strategic airlift of supplies to Berlin's two million inhabitants, began. Great Falls AFB was selected as the replacement training facility for aircrews headed to Germany to fly as part of the Berlin Airlift.

The roots of the 120th Fighter Wing, Montana Air National Guard, began on June 27, 1947, with the activation of the 186th Fighter Squadron. In 2015, the unit became the 120th Airlift Wing, equipped with the Lockheed C-130H Hercules.

Great Falls AFB played a major role in the aerial defense of North America. The 29th Air Division activated at Great Falls in early 1950, bringing with it fighter-interceptor squadrons, aircraft control and warning squadrons, and Ground Observer Corps detachments as a stopgap warning network to watch for Russian bomber penetrations until replaced by radar sites across the northern United States and Canada.

In 1954, the base was aligned under the Strategic Air Command (SAC) and the 407th Strategic Fighter Wing was assigned. The primary mission of the wing's F-84 fighters and KB-29 air tankers was to provide aerial escort for SAC's long-range Convair B-36 Peacemaker bombers. On August 21, 1954, the 407th's vice commander, Col. Axel Malmstrom, died when his Lockheed T-33 Shooting Star crashed two miles west of the Gore Field airport, and the base was renamed for him.

In 1958, the Air Force planned for a squadron of Bombarc surface-to-air missiles to be deployed to permanent air defense sites to defend against bomber attacks, but this was cancelled. The replacement Nike-Ajax and Nike-Hercules sites were also canceled, and the Safeguard anti-ballistic missile site, designed to protect the Minuteman intercontinental ballistic missile (ICBM) complex from a Russian attack, was canceled as well.

The 341st Strategic Missile Wing was activated at Malmstrom AFB on July 15, 1961, and construction started in March 1961. On July 27, launch facility Alpha-09 gained the distinction of being the first Minuteman I ICBM site

Throughout the 1970s and 1980s, the wing's ICBMs remained on alert while undergoing extensive weapon systems upgrades. North American Aerospace Defense Command (NORAD) electronic countermeasures personnel were stationed at Malmstrom AFB. In 1988, the Hardened Missile Mobile Launcher for the Small ICBM concept was tested at Malmstrom to verify its ability to operate off-road and in harsh winter conditions, as expected to be encountered at the Air Force's northern ICBM wings. This was in response to the growing numbers of Russian ICBMs, designed to deliver multiple warheads against SAC missile and bomber bases. This was viewed as a first-strike capability intended to reduce the response from the United States.

A major restructuring of the wing occurred in 1989, when SAC relocated the 40th Air Division to Malmstrom AFB and assigned it to host responsibilities for the 301st Air Refuelling Wing and 341st Strategic Missile Wing. The 301st deployed to Moon Island in the Persian Gulf during Operation Desert Storm. During the deployment, the 301st flew 443 combat sorties refueling 936 coalition aircraft, transferring 33.5 million pounds of fuel. The 341st deployed security, civil engineering, services, and support personnel for Desert Storm.

On June 14, 1991, the 40th Air Division deactivated, returning base host responsibilities to the 341st. On September 1, 1991, the 341st Strategic Missile Wing was redesignated the 341st Missile Wing. Also in 1991, the Strategic Arms Reduction Treaty was officially implemented by Pres. George W. Bush. President Bush ordered all Minuteman II ICBMs taken off alert on September 27, with launch facilities destroyed, missile alert facilities deactivated and offered for sale, and underground launch command centers destroyed and their entrances filled in. In November 1991, the 12th Missile Squadron became the first to have its Minuteman IIs removed in compliance with the president's deactivation order. It would take three and a half years until the final Minuteman II ICBM in SAC's inventory was removed at on August 10, 1995. As the Minuteman II ICBMs were removed, a new program called Rivet Add was launched, modifying the base's remaining 150 Minuteman II launch facilities to accommodate the Minuteman III.

In the early 1990s, host base responsibilities changed several times due to massive Air Force restructuring programs. On January 15, 1992, Malmstrom's host responsibilities were again transferred, this time to the Air Mobility Command's 301st Air Refuelling Wing. On June 1, 1992, responsibility for the ground-based Minuteman IIIs was transferred to Air Force Space Command. The 341st Missile Wing transferred from Air Mobility Command to Air Force Space Command, and the 43rd Air Refuelling Wing was transferred and redesignated as a group, transferring base host responsibilities back to the 341st.

In 1995, the Rapid Execution and Combat Targeting system signaled the first complete overhaul of the Minuteman III's command and control system. The wing also completed the transfer of 120 Minuteman IIIs from Grand Forks AFB's 321st Missile Group. This brought Malmstrom's Minuteman III force to 200 on alert, and was the first time since 1991 that it had a full complement of a single-weapon system.

The 1995 Base Realignment and Closure Commission led to the deactivation of the 43rd Air Refueling Group, transferring its aircraft to the 6th Air Base Wing at McDill AFB, Florida. On January 1, 1997, Malmstrom's runway was declared inactive for the first time in the base's history. Later in 1997, the 819th Red Horse Squadron arrived on the base. It was the first Red Horse Squadron to be comprised of active duty, Air National Guard, and Air Force Reserve. The squadron provides combat engineering personnel and equipment to military and humanitarian operations worldwide on short notice. The latest addition to Malmstrom's mission, the Network Operations Center, performs passive space surveillance and provides users with reliable tracking or near-earth and deep space satellites.

Malmstrom AFB's physical appearance has undergone many changes since 1995. The construction of new buildings and renovation of existing family housing units, dormitories, work facilities, and general support infrastructures transformed the base. These improvements to support the base's mission and quality of life resulted in Malmstrom's being named Best Air Force Installation in the continental United States in 1999.

The wing's Minuteman IIIs have undergone improvements and upgrades. They are subject to constant efforts to keep the missiles safe, secure, and reliable well into the 21st century. These programs include replacement of the aging guidance systems, remanufacture of solid-propellant rocket motors, replacement of standby power systems, repair of launch facilities, and installation of updated survivable communications equipment. These efforts will extend the life of the Air Force's ICBM force, guaranteeing Malmstrom AFB's future.

One

Construction and World War II 1942–1945

Malmstrom Air Force Base traces its history to September 1, 1939, when Nazi Germany invaded Poland, plunging Europe into war, which became a world war after the Japanese naval and air attack on Pearl Harbor, Hawaii, on December 7, 1941. Concern over the war in Europe caused the Great Falls Chamber of Commerce to contact its two US senators—Burton K. Wheeler and James E. Murray—requesting they consider supporting a US Army Air Corps (replaced on June 21, 1941, by the Army Air Forces) base in Great Falls and others in Montana. In 1941, the Civil Aeronautics Authority provided funds for the development of Great Falls's municipal airport. This was renamed Gore Field; today, it is Great Falls International Airport. Great Falls made a significant contribution to the nation's military force fighting the two-ocean war. Gore Field's primary mission was establishing the starting point and operating an air route to Ladd Field at Fairbanks, Alaska, to deliver Lend-Lease aircraft to the Soviet Union for fighting on the Eastern Front against Germany. Aircraft flown by Army Air Forces pilots to Alaska on the Alaskan-Siberian route not only delivered aircraft but 1,717,712 pounds of cargo consisting of aircraft spare parts, tools, miscellaneous equipment, explosives, and medical supplies processed and shipped through Great Falls to Russia. After the end of the war, US intelligence agencies revealed espionage operations by Russian operatives in the United States, with the majority of that information shipped in diplomatic cases on Lend-Lease aircraft through Great Falls. The United States turned a blind eye to these activities because it needed the Soviet Union to keep fighting the Germans and then, after the end of the war in Europe, enter the war in the Pacific against the Japanese. Hundreds of tons of nuclear weapon materials, strategic intelligence reports, and the plans for some of the most advanced aviation, electronic, and heavy industrial technology were transferred through Gore Field and East Base, Great Falls.

A flat, grassy plain six miles east of Great Falls was selected by the Army Air Force survey team as the site for the construction of Great Falls Army Air Force Base, to train Boeing B-17 Flying Fortress bomber crews. Construction began on June 8, 1942. (Courtesy 341st Missile Wing.)

Paving work on Great Falls Army Air Force Base was facilitated by an on-site asphalt plant. The runway was graded by bulldozers and heavy earth scrapers to reach the desired level and incline for runways, taxiways, and parking aprons prior to paving. (Courtesy 341st Missile Wing.)

This aerial view of Great Falls Air Force Base shows the three runways, taxiways, parking apron, and base support buildings. (Author's collection.)

Buildings constructed on Great Falls Army Air Force Base were intended to be temporary, with a life expectancy of five to seven years. Contractors used local wood sources to build these structures to speed construction. They could complete a single-story barracks on concrete block foundations in about one hour. These buildings were not well insulated but offered bearable living conditions for trainees. (Courtesy 341st Missile Wing.)

Pictured are completed hangars with a control tower to direct training operations, especially landings and takeoffs. The hangars were large enough to hold B-17s for maintenance. Later, thousands of Lend-Lease aircraft were modified here for cold weather operations in Russia. (Courtesy Malmstrom Air Force Base Museum.)

A standard 10-man B-17 aircrew consisted of four officers and six enlisted men. This B-17 crew trained at Rapid City Army Air Force Base in South Dakota. The person identified as "me" is Alfred Marcello, a radio operator. Their B-17F is parked behind them. (Courtesy 28th Bomb Wing.)

Training at Great Falls Air Force Base depended greatly on accurate weather forecasts. These two Army enlisted men are manually preparing weather forecast charts for B-17 aircrew pretraining mission briefings. This was critical given the wild swings of Montana weather. (Courtesy 341st Missile Wing.)

After leaving Great Falls and being shipped to England, the trained bomb groups used the same equipment that they trained with stateside. Teletypes received bombing mission strike orders for the bomb group over occupied Europe. The equipment pictured here is in a recreated operations/control tower used in England, built to standard specifications, on display at the National Museum of the US Air Force. (Author's collection.)

B-17 aircrews with the Eighth Air Force based at airfields in England fought at high altitude in below-zero temperatures. Each crew member had to have the equipment to survive and fight at these altitudes, usually 20,000 feet. This included woolen underwear, flight suits, jacket, and pants, and insulated flight boots and gloves, which made firing .50 caliber machine guns and operating equipment and flying the aircraft difficult. Each wore a heavy leather flight helmet, over which was a steel combat helmet, along with a flak vest for protection against German 88-mm flak artillery. Each wore flight goggles and oxygen mask, a parachute, and a Mae West life preserver for over-water flights from and to England. This gunner outfit is on display at the National Museum of the US Air Force. (Author's collection.)

Women work on maintenance and inspection of a Bell P-63 King Cobra. The work also included winterization prior to the long flight to Alaska, where aircraft were turned over to Soviet air force pilots for transit flights through Siberia to the Eastern Front. (Courtesy 341st Missile Wing.)

Women worked in all types of jobs at Great Falls Army Air Force Base during Lend-Lease operations to prepare aircraft for transfer to the Soviet Union. Here, a woman worker spray paints a red star on a C-47's fuselage. Such work was done outside whenever possible, speeding modifications and preparations to turn aircraft over to the Soviet Union. (Courtesy 341st Missile Wing.)

Male workers, civilians, and military were part of Lend-Lease aircraft preparations at Great Falls Army Air Forces Base. This enlisted man is spray painting a red star on the underside of a C-47's wing inside one of the hangars. (Courtesy 341st Missile Wing.)

Shown is the interior of enlisted barracks on Great Falls Army Air Force Base during World War II. Double steel frame bunk beds line both sides of the barracks, leaving a center aisle. Four wood- or coal-burning pot belly stoves heat the barracks. Someone was assigned stove watch to keep the stoves running 24 hours a day during cold periods when temperatures could drop well below zero. Some of the men pictured here are wearing heavy, long winter overcoats inside the barracks. Wood cross beams and trusses supported heavy snow loads. (Courtesy 341st Missile Wing.)

The flight line was very busy during World War II after the base shifted from B-17 training operations to assisting Russia with aircraft and supplies to fight the Germans on the Eastern Front. Pictured are B-26s, B-46s, and B-47s. (Courtesy 341st Missile Wing.)

In the flight operations area, at the control desk, personnel were assigned to outgoing scheduled flights, including pilots, aircrews, and others transiting through the base to other stateside military bases. Arrivals and departures were updated by hand with chalk on the debarkation boards. This was the heart of base transportation operations. (Courtesy 341st Missile Wing.)

Army Air Force pilots pose in front of a P-39 on the parking ramp at Great Falls Army Air Force Base. Military pilots flew many different types of Lend-Lease aircraft to Great Falls Army Air Force Base, support by WASP pilots, but only the men flew the aircraft north through the wilds of western Canada and the rugged areas of Alaska to Ladd Airfield. Once the aircraft had been delivered, they flew back to Great Falls in a C-47, ready to repeat the process. These Lend-Lease flights were dangerous. If an aircraft crashed in the wilderness, especially during winter, odds of survival were poor. (Courtesy 341st Missile Wing.)

North American B-25 Mitchell twin-engine, medium bombers arrived in Great Falls on railroad flatcars, in heavy wood crates. They were loaded by crane onto flatbed tractor-trailers for the trip to Great Falls, where they were assembled and modified for cold weather operations in Russia. (Courtesy 341st Missile Wing.)

P-39s on Great Falls Air Force Base wait for ferry flights to Alaska and turnover to Russian pilots for the long flight through Siberia to the Eastern Front. The Russian air force liked the rugged P-39. (Courtesy 341st Missile Wing.)

A Soviet air force officer is pictured with an A-26 bomber at Ladd Airfield, Alaska. Russian officers inspected each aircraft even though they had been inspected and cleared at Great Falls. They refused any aircraft that did not pass muster. These aircraft had to be repaired, often because of worn tires, which the Russians demanded to be new before the flight over Siberia. Once repaired, re-inspected, and accepted, Soviet pilots flew the aircraft to the Eastern Front. (Courtesy National Museum of the US Air Force.)

Weather at Great Falls could be as brutal and unforgiving as that encountered by aircrews on the route to Alaska. Under the tail of a B-25, another B-25 is visible to the right on the snow-covered, wind-swept parking ramp. There are two portable engine heaters positioned near each B-25. These heaters were necessary to warm the thick oil and fuel so that the engines could be started without damage. Heaters are also positioned around each C-47. Their wings are covered with snow, which accumulated on the metal parts of the aircraft. Flight line operations during cold winter temperatures were dangerous, with frostbite occurring within minutes to exposed skin. Maintenance personnel could not touch the aircrafts' aluminum skin without freezing to the bare metal. (Courtesy 341st Missile Wing.)

Lend-Lease aircraft, after modifications and inspections, were test flown to determine if they were ready for the flight to Alaska. Many test flights were conducted in extreme conditions. If the flight test was acceptable, the aircraft was certified for flight to Alaska. (Courtesy 341st Missile Wing.)

Russians in downtown Great Falls were not a rarity for the city's citizens. They shopped in town, making purchases in US currency or gold certificates. Most of these gold certificates predated the Communist revolution, taken from Tsarist banks in 1918. The Russians' purchases were boxed and loaded onto a C-47 bound for Russia with diplomatic protection, exempt from US inspection. (Courtesy Great Falls Museum.)

A restored World War II P-39 is pictured in front of Lend-Lease Hangar No. 2 at Great Falls International Airport in 2015 during a commemorative fly-in. This was the launching point for a commemorative flight from Great Falls to Alaska, through Siberia, and on to Moscow. A C-47 was delivered and turned over to an aircraft museum in Moscow, recording the significance of Lend-Lease aircraft to Russia during World War II. (Courtesy Great Falls International Airport.)

Seen here is the rear of Hangar No. 2; with these hangar doors permanently secured in the closed position, window panels remain intact. The side windows are long since gone, covered with plywood sheeting. The Great Falls International Airport's perimeter road runs behind the hangar. (Author's collection.)

Pictured is Hangar No. 1's west side. The hangar's large interior space allowed multiple aircraft to be serviced, and its ceiling beams offered lift support for aircraft repair. It was well insulated for year-round operations. (Author's collection.)

Two

Berlin Airlift and the Start of the Cold War 1948–1949

In February 1948, the Communists seized power in Czechoslovakia, and in June, the Soviet Union tried to sever Berlin from the West. At the end of World War II, the victorious Allies had reached an agreement about the division of Germany into three zones of occupation. The British, French, and American zones became the Federal Republic of Germany (West Germany), while the Soviet Union occupied the eastern zone, eventually becoming the German Democratic Republic (East Germany). Berlin lay within the Soviet zone, itself divided into zones of occupation by the four former allies. The agreement assured access to the city along ground and air routes from the West through Soviet-controlled East Germany. The Soviet Union reneged on this agreement in June 1948, closing overland routes into West Berlin. Communist demonstrators drove out elected members of the combined city assembly. For the next few months, various functions of the Berlin city government were divided between east and west. Wartime cooperation was rapidly breaking down. The US Air Force selected Great Falls Air Force Base as the replacement aircrew training site for Berlin Airlift–bound Douglas C-54 Skymaster aircrews, activating the 517th Air Transport Wing. Great Falls would have been the last place with any connection to the Berlin Airlift, but this was, in fact, where many recalled World War II pilots came for training before taking part in the operation. Hundreds of pilots and crewmen were checked out on the C-54 Skymaster and on flight procedures on mock-up duplicates of facilities at Rhein-Main, Wiesbaden, and Berlin. Great Falls proved to be an ideal location for the C-54 replacement training unit. The weather was similar to that in Germany, and planners made every effort to duplicate conditions those flying the airlift would face.

This plaque is on display at the Dyess Air Force Base memorial airpark in Texas. (Author's collection.)

Rapid City, South Dakota, holds a significant place in the history of Cold War relations between the United States and the Soviet Union. The tearing down of the Berlin Wall separating East and West Berlin started the fall of the former Soviet Union. Two sections of the concrete wall are on display in Rapid City, south of the civic center. Those who flew the Berlin Airlift set in motion the eventual removal reunification of Germany. (Author's collection.)

Military Air Transport Service (MATS) personnel are loading into a C-54 at Malmstrom Air Force Base for the flight to Germany to support the Berlin Airlift. (Courtesy 341st Missile Wing.)

C-54s became a frequent and constant aerial presence over Great Falls during their training flights in preparation of assignment to Germany and Berlin Airlift flights. The World War II buildings on the base remained operational during training, filled with aircrews, maintenance, and support personnel. (Courtesy 341st Missile Wing.)

The Douglas C-54 Skymaster four-engine MATS transport became the backbone of Berlin Airlift flights due to its considerable cargo capacity of 10,000 pounds versus that of a C47's 6,000 pounds, at a higher speed, allowing more flights per day. This C-54 is about to touch down at Great Falls Air Force Base after completing a training flight. High, rugged terrain is visible to the south of the base. (Courtesy 341st Missile Wing.)

A Douglas C-54 Skymaster is undergoing scheduled maintenance. Technicians are working on one of the aircraft's engines, assisted by maintenance steps in front of the transport's wings. (Courtesy 341st Missile Wing.)

At Tempelhof in West Berlin, a German worker poses in front of US garbage cans. To the left is a can for edible garbage, which West Berliners searched for food. (Courtesy 341st Missile Wing.)

MATS personnel line up for loading into a Douglas C-54. Women played a vital support role in the Berlin Airlift. (Courtesy 341st Missile Wing.)

Sandbags were used to give aircrews hands-on experience handling heavily loaded Skymasters. Each aircrew had to make three landings at 70,000 pounds gross weight before graduating from the training. The three-week course at Great Falls Air Force Base provided preflight and flight training for all crew members, supplementary ground training for pilots, and on-the-job training for flight engineers. (Courtesy 341st Missile Wing.)

Maintenance personnel received extensive and specialized training on C-54 engines, vital for Berlin Airlift and Great Falls Air Force Base training operations. A fully functioning C-54 engine repair and overhaul shop was used not only for training aircraft mechanics, but also for keeping as many of the training Skymasters operational as possible. (Courtesy 341st Missile Wing.)

A considerable number of women military personnel, officer and enlisted, were assigned to West Germany to support Berlin Airlift operations. Silver braid on the flight caps of two personnel signify that they are officers. They and others are being processed for pre-boarding aircraft to West Germany. (Courtesy 341st Missile Wing.)

This model of a Douglas C-54 transport flown by American aircrews during the Berlin Airlift is on display at the Malmstrom Air Force Base Museum. (Author's collection.)

This Douglas C-54 crashed on the Malmstrom Air Force Base runway during a training flight. (Courtesy 341st Missile Wing.)

Three

Ground Observer Corps and Air Defense 1949–1976

After Pres. Harry Truman's announcement of a Soviet atomic detonation in the summer of 1949, the Air Force created a stopgap method to detect penetrating Russian bombers until a radar network was in place, the Ground Observer Corps (GOC), under an Air Force program called Operation Skywatch. The number of civilian volunteers necessary for the GOC was estimated at 500,000, with 200,000 enrolled by February 1952, operating 49 filter centers and 20,000 observation posts. If an unidentified aircraft was spotted, GOC observers passed this information by telephone to GOC officials and then on to the Semi-Automatic Ground Environment (SAGE) system. Many Montana residents did not believe there would be a nuclear war with the Soviet Union, but attitudes rapidly changed during and after the Cuban Missile Crisis and the fear of living through a nuclear attack by the Soviet Union. The 4773rd Ground Observation Squadron at Malmstrom operated from April 16, 1951, to March 25, 1959.

On December 18, 1953, the 407th Strategic Fighter Wing activated at Great Falls Air Force Base (inactivated on July 1, 1957), with the 407th Air Refueling Squadron operating the Boeing KB-29 Superfortress tankers, along with the 515th, 516th, and 517th Strategic Fighter Squadrons, with the Republic F-84 Thunderjet fighter-interceptor. The base was renamed for Colonel Malmstrom, killed in a T-33 crash in 1956. The 91st Strategic Reconnaissance Squadron experimented with the Fighter Conveyance system. On October 1, 1955, the 407th Strategic Fighter Wing converted to the F-84F. SAC activated the 4061st Air Refueling Wing, 407th Air Refueling Squadron, converting from the KB-29 to the Boeing KC-97 Stratofreighter tanker. The 29th Fighter-Interceptor Squadron also flew the F-94 Starfire, F-89 Scorpion, F-101B Voodoo, and F-89Js, armed with conventional air-to-air missiles and two Genie nuclear-armed air-to-air missiles. The 71st Fighter-Interceptor Squadron next flew the Convair F-106 Delta Dart. On July 1, 1974, the 17th Air Defense Systems Evaluation Squadron flew the EB-57 Canberras to test Air Defense Command installations, especially radar sites.

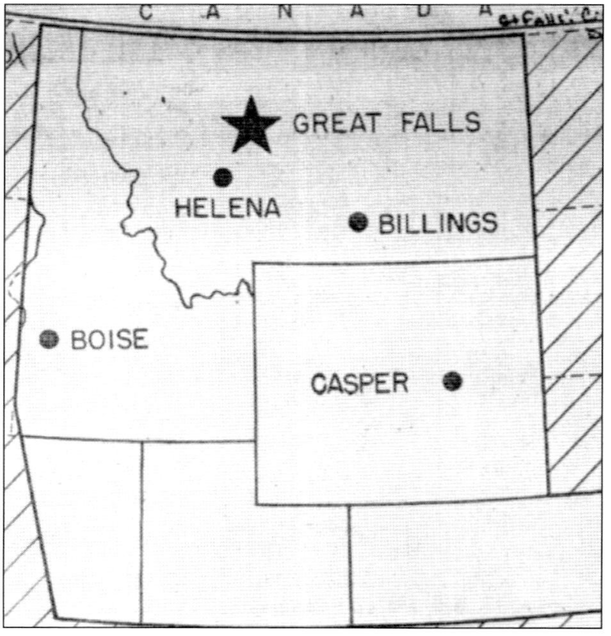

Great Falls and Malmstrom Air Force Base were the northern hub of the nation's important air defense operations to protect against Russian bombers. A network of GOC posts and four filter centers operated in Montana, Idaho, and Wyoming, with the large concrete SAGE building on Great Falls Air Force Base, the hub for Operation Skywatch. The 4773rd Ground Observer Squadron, a unit of the 29th Air Division on the base, served as headquarters for the scattered observer posts, with sightings reported to the appropriate filter centers. (Courtesy Montana State Historical Center Archives.)

Large numbers of civilian volunteers were needed to man Montana's GOC posts, which were considered to be in the primary Russian bomber corridor over the North Pole into the United States. Recruiting posters were placed in public buildings as during World War II's largely successful war bond drives. (Courtesy Montana State Historical Society Archives.)

The Air Force prepared this Ground Observer's Guide for the hundreds of thousands of GOC volunteers. It contained photographic guides and silhouette keys for identifying US and Soviet military aircraft. Montana GOC volunteers quickly became experts at identifying US aircraft frequently seen over the ground observer post. Many sightings were civilian aircraft off course in the open areas of Montana. (Courtesy Montana State Historical Society Archives.)

A GOC post could be as simple as a volunteer's backyard deck or patio, with an aircraft recognition chart, binoculars to scan the horizon, a worksheet or log to record sightings, and a telephone to call the filter center. More elaborate watchtowers, similar to those used in Montana forests to watch for fires, were also built. (Courtesy Montana State Historical Society Archives.)

In an effort to raise volunteers for the GOC, the Air Force used large billboards at major highway intersections to inspire patriotism. The Air Force started the GOC with only 40 percent of the volunteers needed. (Courtesy 341st Missile Wing.)

This is a filter center plotting board for tracking aircraft positions from GOC posts, as done during World War II. This was a makeshift operation, before the use of overhead screens and computers, but it worked. (Courtesy Montana State Historical Society Archives.)

The *Great Falls Tribune* printed this ad for an underground bomb shelter, a step up from basement bomb shelters. It was made of steel, designed to protect a family from an atomic blast and radiation, with a secure entrance that could be locked from the inside. It was not cheap, and three-year loans were available to help pay for them. (Courtesy Great Falls Museum Archives.)

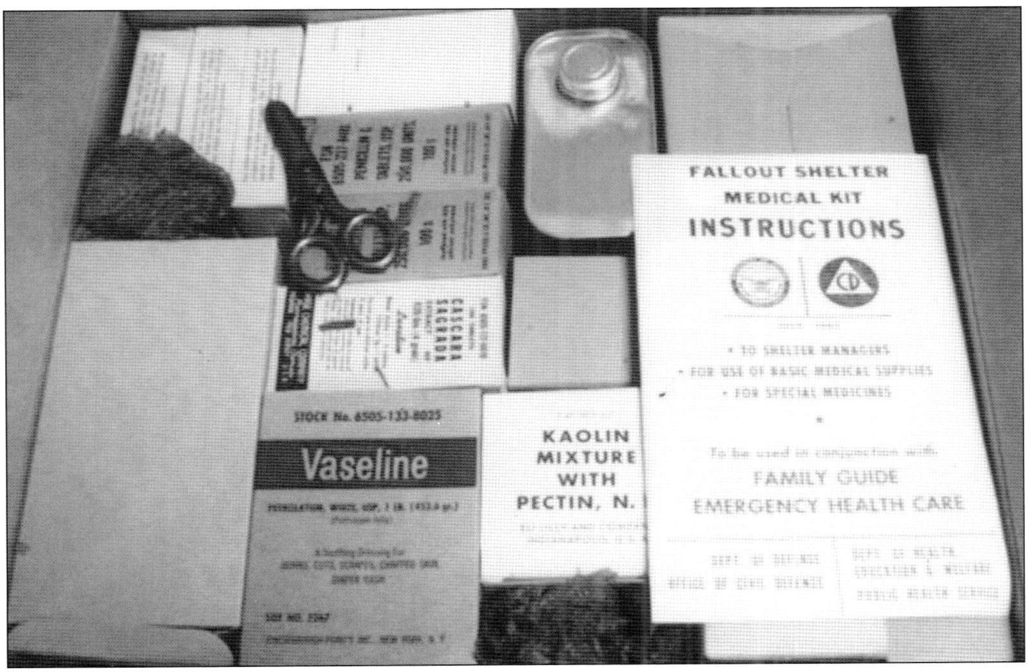

As relations with the Soviet Union deteriorated, the new media of television was used to urge Montanans to undertake civil defense. One request was for citizens to prepare a fallout shelter medical kit, since medical care after a nuclear attack would probably be nonexistent for a long time. (Courtesy Great Falls Museum Archives.)

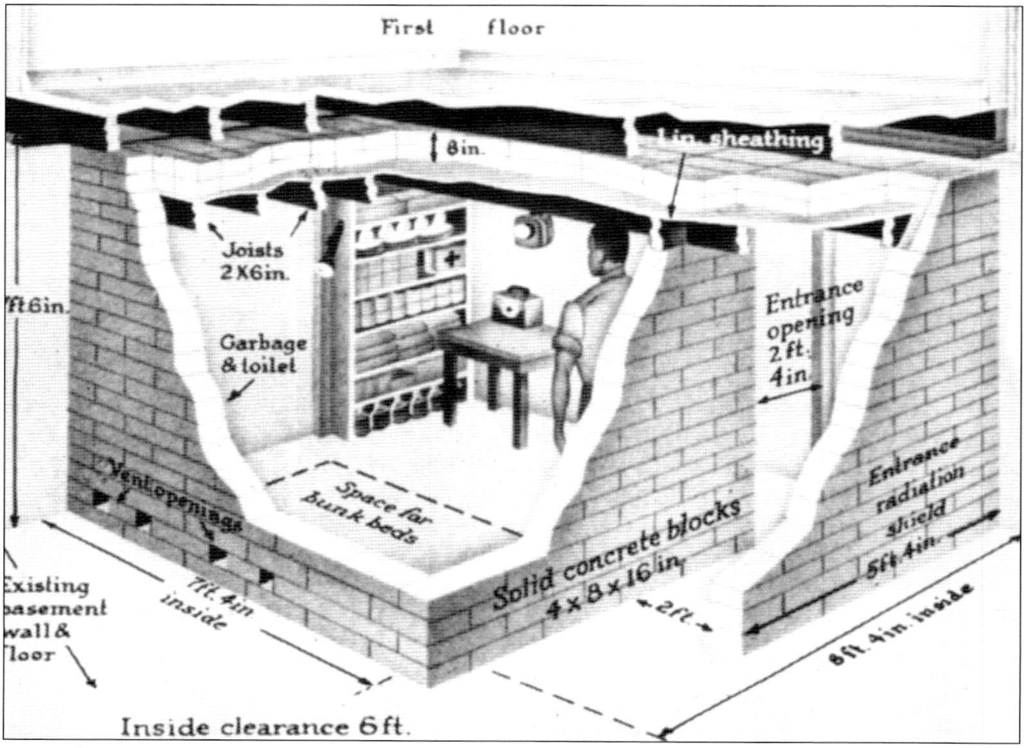

As nuclear war between the United States and the Soviet Union appeared imminent, the need for a family fallout shelter surfaced with advertisements for shelters for protection outside cities with a civil defense shelter program. Northwest corners in concrete basements were turned into bomb shelters. The fear of nuclear attack increased during the events leading up to the Cuban Missile Crisis. Many farmers and ranchers with root cellars modified them into shelters. (Courtesy Great Falls Museum Archives.)

In January 1965, there were 603 shelter facilities, but only 414 were marked, and only 376 stocked with supplies, providing a total of 235,518 approved individual berths. Shown is a family fallout shelter. Over 66 percent of the state's population had no such protection. (Courtesy Great Falls Museum Archives.)

Pictured is the Convair F-106 Delta Dart, a supersonic fighter-interceptor on Air Defense Command alert. This F-106 is intercepting a Russian Tu-95 bomber. (Courtesy 341st Missile Wing.)

Pictured is a Republic F-84 straight-wing Thunderjet parked on the flight line adjacent to the Great Falls Air Force Base control tower. (Courtesy 341st Missile Wing.)

A 407th Air Refueling Squadron Boeing KB-29 Superfortress aerial tanker is seen here. It was equipped with the first model of the Boeing Flying Boom. The tanker is refueling a Republic F-84 Thunderjet fighter-interceptor using the World War II design. It was used by the SAC as a long-range fighter escort, primarily for the Convair B-36 bomber. (Courtesy Strategic Air Command.)

This Republic F-84F Thunderstreak is on display at the South Dakota Air and Space Museum. The F-84F was a fighter-interceptor intended to intercept Soviet bombers. (Author's collection.)

The base was named for Col. Einar Axel Malmstrom, deputy commander, 407th Strategic Fighter Wing, after he was killed in a Lockheed T-33 Shooting Star crash. (Courtesy 341st Missile Wing.)

This Lockheed T-33 Shooting Star is on display at the Malmstrom Air Force Base Air Park. (Author's collection.)

This bronze plaque honors Colonel Malmstrom. It is on display in the Malmstrom Air Force Base memorial park. (Author's collection.)

Pictured is the Convair B-36D with six piston pusher engines and two turbojet engines at the end of each wing, outboard of the piston engines. The B-36D was referred to as "six turning and four burning." It is on the ramp at Andersen Air Force Base, Guam. This was the first B-36D to land at Andersen. Its great size is apparent relative to the personnel and vehicles around the bomber. (Courtesy 28th Bomb Wing.)

Pictured is a B-36D landing at Malmstrom Air Force Base for Fighter-Conveyance (FICON) feasibility tests. (Courtesy Strategic Air Command.)

Initial FICON tests used the straight wing DF-84 to launch and recover the GRB-36D. The specially designed RF-84Ks would be ferried close to the reconnaissance target, launched from the B-36 at high altitude, make a high-speed reconnaissance pass over the target, return to the mother aircraft in international airspace, re-hook, and be ferried back to its home base and released to land, followed by the B-36. (Author's collection.)

Pictured is the RF-84K on display at the National Museum of the US Air Force in the Cold War Gallery. The FICON hook is mounted in front of the cockpit. (Author's collection.)

The 4061st Air Refueling Wing flew the Boeing KC-97G Stratofreighter tanker. This is a full view of a KC-97G tanker, with the Boeing Flying Boom. On the right wing is an outboard auxiliary fuel tank and a J-57 turbojet engine to add speed so the tanker could refuel turbojet fighters and bombers. (Author's collection.)

The 29th Fighter-Interceptor Squadron upgraded to the two-place, F-101B Voodoo high-performance supersonic interceptor. (Author's collection.)

The 29th Fighter-Interceptor Squadron replaced its F-101s for F-89Js, armed with conventional air-to-air missiles and two Genie nuclear air-to-air missiles. It is on display alongside the entrance to the 120th Airlift Wing at Great Falls International Airport. (Author's collection.)

Shown is a head-on view of a Convair F-106 Delta Dart on display at the 123rd Airlift Wing Air Park (formerly the 120th Fighter Wing) outside the Montana Air National Guard entrance. (Author's collection.)

Four

Air Defense Radars and Aerial Refueling 1953–1971

Great Falls Air Defense Sector was created in March 1959, assuming control of former Air Defense Command Central Defense Force units with a mission to provide a radar barrier across the northern states, and the SAGE system computer-based command center. It operated from 1962 to 1976, controlling fighter-interceptor units at air bases in the area. The huge center required a force of 5,000 NORAD personnel housed in a four-story, windowless concrete blockhouse large enough to contain two massive computerized command and control AN/FSQ-7 computers, each with 50,000 vacuum tubes. The heat was a problem, requiring a powerful cooling system. Even with this technology, SAGE centers maintained an incredible near 100 percent reliability.

To enhance air defense, the base housed SAC strategic aerial tankers. The 407th Strategic Fighter Wing initially operated KB-29P tankers, a modification of the World War II Boeing B-29 Superfortress. The 4061st Air Refueling Wing, 97th Air Refueling Squadron, flew aerial refueling support missions for SAC from Malmstrom Air Force Base until July 1961, upgrading to piston-powered Boeing KC-97 Stratotankers. In 1956, SAC started phasing out the KC-97, replacing it with the Boeing four-turbojet KC-135A Stratotanker, which could match the speed of turbojet fighters and bombers. In October 1992, the 97th Air Refueling Squadron was equipped with re-engined KC-135R tankers, which provided more engine efficiency and increased fuel offload capacity. The squadron continued to support deployed tanker task forces throughout the world, including periodic support for the 1700th Strategic Wing (Provisional) in Riyadh, Saudi Arabia. It also participated in Operation Restore Hope in Somalia. The tankers were withdrawn from Malmstrom Air Force Base with the inactivation of the 301st Air Refueling Wing.

Pictured is the SAGE Direction Center (DC-20), originally used for the air defense of the north-central United States. The four-story, windowless concrete blockhouse was built specifically to house two large AN/FSQ-7 computers, using thousands of vacuum tubes to conduct the necessary computations tracking incoming airborne threats and other unidentified aircraft. (Author's collection.)

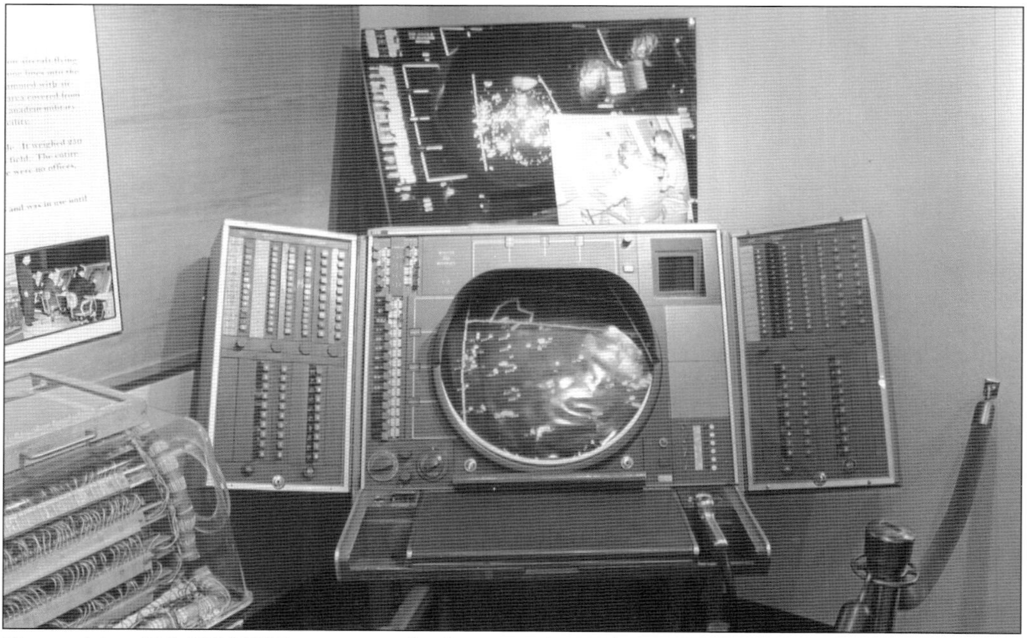

Pictured is a QA-1008/FSG situation display console, with a 19-inch diameter cathode ray tube (CRT) screen, on display at the Malmstrom Air Force Base Museum. (Author's collection.)

Shown is one of the many magnetic drums required to operate the SAGE computer. These drums were the precursors of the 1960s-designed floppy disk drives. The drums were for buffer storage of input and output data, system status data, and computer programs. There were a total of 12 used on one computer, providing data storage of 150,000 thirty-two-bit words. This one is on display at the Malmstrom Air Force Base Museum. (Author's collection.)

Seen here is one of the many computer cabinets with rows of vacuum tubes. Vacuum tubes created a lot of heat, requiring cooling to keep them from overheating and bursting, destroying the computer. This one is on display at the Malmstrom Air Force Base Museum. (Author's collection.)

Malmstrom Air Force Base's SAGE building was packed with two massive computers and radar consoles. Pictured is one of the approximately 90 QA-1008 FSQ situation display consoles. At this time, the console was considered advanced technology. (Courtesy 341st Missile Wing.)

Pictured is a model of a Martin EB-57 Canberra twin-engine turbojet electronics evaluation, jamming, and faker aircraft. They were stationed at Malmstrom Air Force Base to test NORAD's radar capabilities and were used worldwide to test radar warning capabilities. The model is on display at the Malmstrom Air Force Base Museum. (Author's collection.)

This is a view of a Martin EB-57 Canberra on display at the Malmstrom Air Force Base Air Park. (Author's collection.)

Pictured is the left-side, underwing pylon mounted electronics pod for the EB-57 Canberra's evaluation missions. Other EB-57s are visible on the opposite side of the aircraft ramp. (Author's collection.)

This Martin EB-57 Canberra is on display at the South Dakota Air and Space Museum at Ellsworth Air Force Base. It was flown against Air Defense Command radar sites to determine their detection capabilities. (Author's collection.)

This Nike-Ajax surface-to-air missile was used in area defense of US military and civilian targets against Russian bombers carrying nuclear bombs. It is on display at the South Dakota Air and Space Museum. (Author's collection.)

Pictured is a Boeing KC-97 Stratotanker. It was the first KC-97 aerial tanker, marking the beginning of the transition of Great Falls Air Force Base to headquarters for the 4061st Air Refueling Wing. The first KC-97 on the base was named *City of Great Falls*. (Courtesy Malmstrom Air Force Base Public Affairs.)

These airborne Boeing KB-29 Stratotankers show the Boeing Flying Boom developed for aerial refueling. (Author's collection.)

This Boeing KC-97 tanker is on display at the Malmstrom Air Force Base Museum. This type of tanker was the first manufactured in large numbers, becoming the backbone of the Strategic Air Command's nuclear bomber refueling. (Author's collection.)

A KB-29 refuels an F-84F. The Boeing KB-29 tanker was a modified B-29 Superfortress to test the flying boom refueling system. The F-84F was initially used by the SAC to provide fighter escort of its Convair B-36 Peacemaker bomber. It needed aerial refueling to stay with the B-36. (Courtesy Malmstrom Air Force Base Public Affairs.)

Pictured is the KC-97 tanker *City of Great Falls* and a flight crew on the ramp at Great Falls Air Force Base. (Courtesy 341st Missile Wing.)

Maintenance of a KC-97's piston engines was a highly technical task. The four piston engines were the same as were used on the B-50 bomber. (Courtesy 341st Missile Wing.)

If a KC-97 crashed with a full load of aviation fuel for the tanker's four engines and a separate load of aviation JP-4 jet fuel, a fire could be disastrous. This KC-97 slid off the edge of the taxiway, heavily damaging the aircraft and starting a fire, extinguished by base fire personnel. (Courtesy 341st Missile Wing.)

A KC-97 refuels a straight-wing F-84 Thunderjet. (Courtesy 341st Missile Wing.)

Shown is the pilot's left-hand cockpit position inside a KC-97 tanker. He flies the aircraft to complete the air refueling. (Courtesy 341st Missile Wing.)

The SAC operated a fleet of 737 KC-97s. Boeing celebrated the production of the 500th KC-97 with a special marking on the aircraft's nose. (Courtesy 341st Missile Wing.)

Pictured is a Boeing KC-135R Stratotanker. The turbojet-powered Stratotanker was developed to match the higher performance turbojet fighters and bombers, which the piston-powered KC-97 could not match. (Courtesy 341st Missile Wing.)

This KC-135R Stratotanker is on display at the Ellsworth Air Force Base ramp during the Thunder Over Dakota 2009 airshow and open house. (Author's collection.)

Five

ANTI-BALLISTIC MISSILE DEFENSE SITE 1968–1974

In 1968, Malmstrom AFB was selected as a site for a Safeguard anti-ballistic missile (ABM) installation to protect the base's Minuteman intercontinental ballistic missiles and supporting facilities. During congressional debates and votes throughout the summer of 1969, little construction activity took place at Malmstrom. However, once funding appeared to be assured, the Army publicly announced the pending deployment of a Safeguard ABM system to sites near Shelby and Conrad. Plans called for the perimeter acquisition radar (PAR) to be located 50 miles north of Great Falls. The missile site radar (MSR) was to be located north-northwest of Great Falls, seven miles southeast of Conrad. By the end of May 1972, the project was about 10 percent complete, and the contractor mobilized resources to launch full-scale construction. However, the project did not make significant headway. On May 26, 1972, Pres. Richard Nixon and Russian general secretary Leonid Brezhnev signed the ABM Treaty, limiting each nation to one ABM site to protect strategic forces and one site to protect its national command authority. With the Grand Forks complex nearly 85 percent complete, the secretary of defense ordered construction suspended at the Malmstrom ABM complex. Its status remained uncertain until October, when the US Senate ratified the Strategic Arms Limitation Treaty (SALT) I and ABM pacts. Terminating the large ABM project became nearly as complex as starting the ambitious project. Approximately 56 contracts had to be reviewed and closed out. Hundreds of compensation claims from suppliers had to be settled. On September 11, 1973, the Huntsville division let two contracts to perform restoration work at the Malmstrom ABM sites. William Clairmont Inc. Bismarck, North Dakota, received a contract to remove the MSR, with Peter Kiewit Sons' Company and Associates assigned to demolish the PAR site. Salvageable equipment and material were removed, and by July 1974, most of the ABM facilities were buried under tons of earth, planted with prairie grasses to blend into the surrounding terrain. After the demolition, only the first level of the unfinished PAR building remained visible.

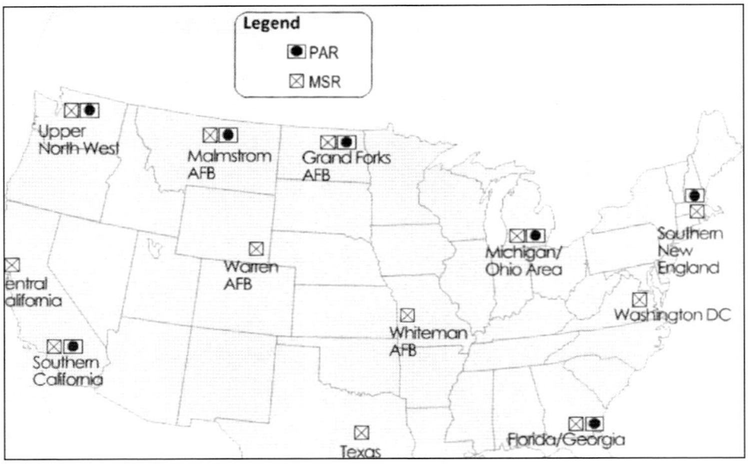

Pictured is the location of the planned sites for the nation's anti-ballistic system sites to provide maximum protection from Russian ICBMs. (Courtesy 341st Missile Wing.)

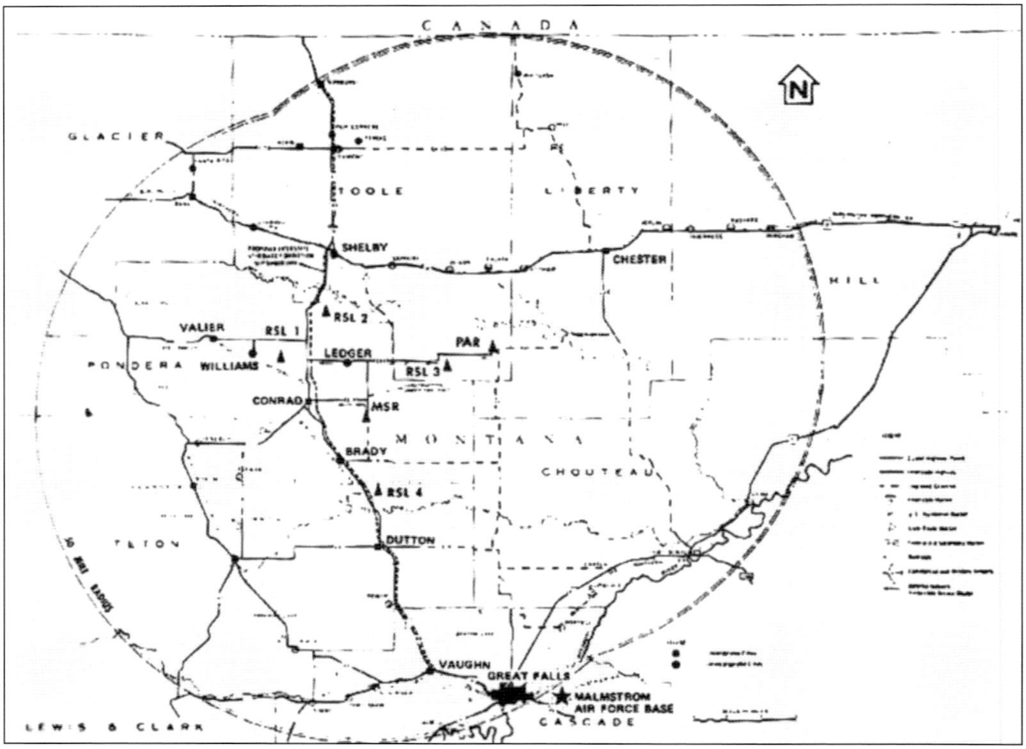

Pictured is the location of Malmstrom AFB's Safeguard ABM installations, including the PAR, MSR equipped with the long-range Spartan and short-range Sprint anti-ballistic missiles, and four remote site launchers—sites 1 through 4—equipped with Sprint anti-ballistic missiles. The complex was north of the 341st Missile Wing's Minuteman squadrons to shoot down incoming Russian missiles. (Courtesy Strategic Air Command.)

Pictured is the first-floor concrete support construction on the PAR site 50 miles north of Malmstrom AFB and 26 miles east of Ledger, Montana. Its super-hardened reenforced concrete walls were designed to withstand a nearby blast from a thermonuclear warhead, but made it cost-prohibitive to break apart and remove after construction was canceled. (Author's collection.)

In 2016, the interior of the unfinished PAR facility was heavily marked with graffiti, piles of garbage, and signs of open fires. The thick concrete floor supported concrete pillars to hold up the upper structure's floors that created the radar building. It was designed to keep on functioning unless it received a direct hit by a nuclear warhead. The foundation and first floor, with the second floor's surface, were completed prior to project cancellation. (Author's collection.)

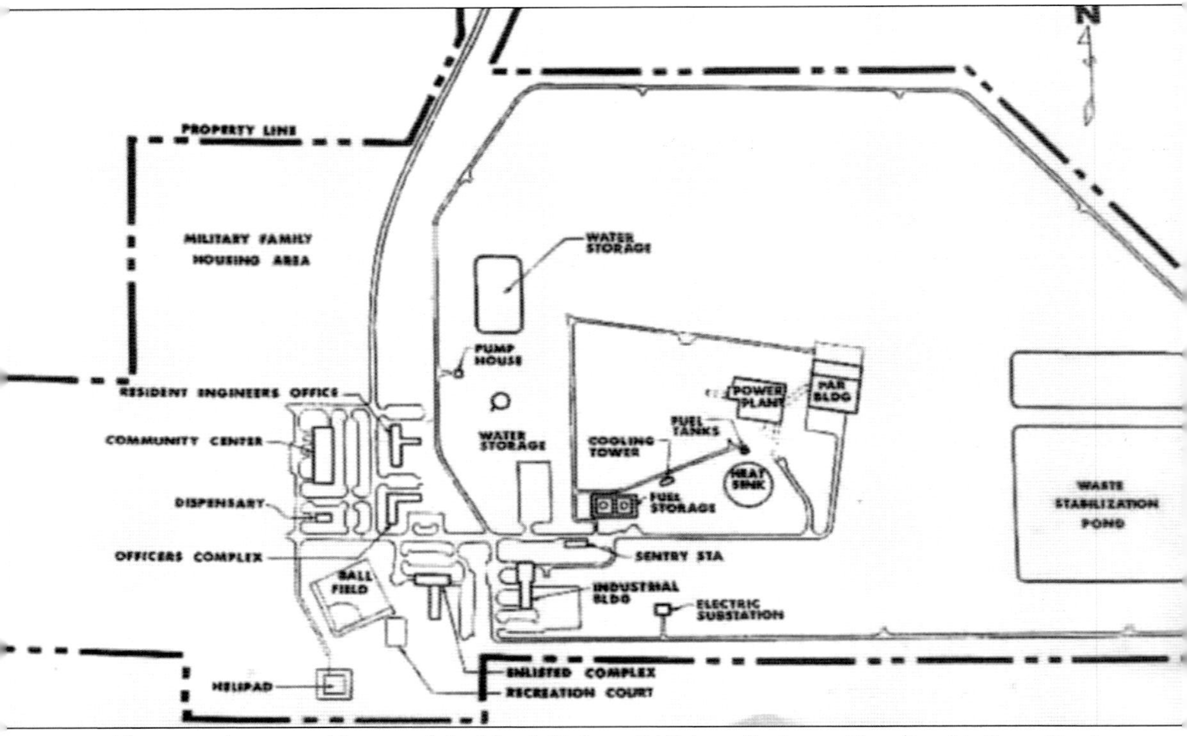

Shown is the general layout of the North Dakota PAR installation at Cavalier Air Force Station. (Courtesy Department of Defense, Department of the Army.)

The administration area at the Safeguard ABM complex is now under private ownership. All electronic equipment was removed, along with the missiles and any salvageable military equipment. The large radar housing called the "pyramid" was not demolished due to its tremendous strength. (Author's collection.)

The only completed and operational PAR still remains as an active duty radar facility at the Cavalier Air Force station, tracking earth-orbiting space debris and satellites. The United States was allowed to retain this radar under the terms of the START treaty, since it was no longer operating as part of an active anti-ballistic missile radar installation. (Author's collection.)

The MSR, nicknamed the "pyramid," held the guidance radars for the Spartan and Sprint anti-ballistic missiles. After the site was deactivated, all salvageable equipment was removed. The complex was offered for sale under the federal bid process as surplus property. Malmstrom Air Force Base was to have a similar installation, but it was canceled. (Author's collection.)

These anti-ballistic missile hardened silos at the northern edge of the remote missile complex contained 30 long-range Spartan and 18 short-range Sprint missiles. Each consisted of a launch chamber, exhaust duct, launch preparation vault, mechanical and electrical equipment vault, and communications antenna. (Author's collection.)

This is a view of the entrance of Remote Launch Site No. 3. The tall concrete structure is the operations building's air exhaust for the auxiliary diesel engines for power production, and the shorter structure is the air intake. The entrance is the small concrete structure at ground level at right center. (Author's collection.)

Pictured is the separate security area that housed the Sprint anti-ballistic missiles in hardened underground vertical launch silos. (Author's collection.)

Shown is the bunkered entrance to the underground remote launch site operations building. The entrance faces away from the expected direction of a nuclear blast from Russian ICBMs coming from the north-northwest. (Author's collection.)

This on-site security entrance led to the remote launch site's operations building and the separately secured Sprint anti-ballistic missile launch area. (Author's collection.)

Six

Minuteman ICBM
1962–Present

Malmstrom Air Force Base's 341st Missile Wing was designated Strategic Air Command's Missile Wing I. In late July 1962, the first Minuteman I arrived at Malmstrom, placed into launch facility Alpha-09. The timely arrival of additional missiles played a critical role in the nation's defense, especially as the Soviet Union's political leadership, led by Premier Nikita Khrushchev, attempted to sneak intermediate-range ballistic missiles (IRBMs) and medium-range ballistic missiles (MRBMs) armed with nuclear warheads onto Cuba in the fall of 1962. On October 24, 1962, SAC accepted the first flight of 10 Minuteman I ICBMs, bringing them up to nuclear deterrent alert status on October 26. Interestingly, two days later, Khrushchev agreed to remove the IRBMs, MRBMs, nuclear warheads, and IL-28 Beagle light bombers with their nuclear bombs from Cuba, transporting them back to the Soviet Union. Minuteman ICBMs are considered a deterrent to a nuclear attack on the United States. Testing of the aging Minuteman III missile is essential to determine the system's reliability as well as give select missile crews the opportunity to launch a missile. A team from the 341st Missile Wing launched an unarmed Minuteman III equipped with a test reentry vehicle on September 5, 2016, at 2:10 a.m., from Vandenburg Air Force Base. The reentry vehicle, with its telemetry package, flew a programmed flight path of 4,200 miles, impacting in the target area at Kwajalein Atoll, Marshall Islands. The test launch was to verify the accuracy and reliability of the Minuteman's systems, providing valuable data to ensure a continued safe, secure, and effective nuclear deterrent. The Air Force has projected the estimated end of life for the Minuteman III as 2030, with a replacement hopefully to be deployed by that date.

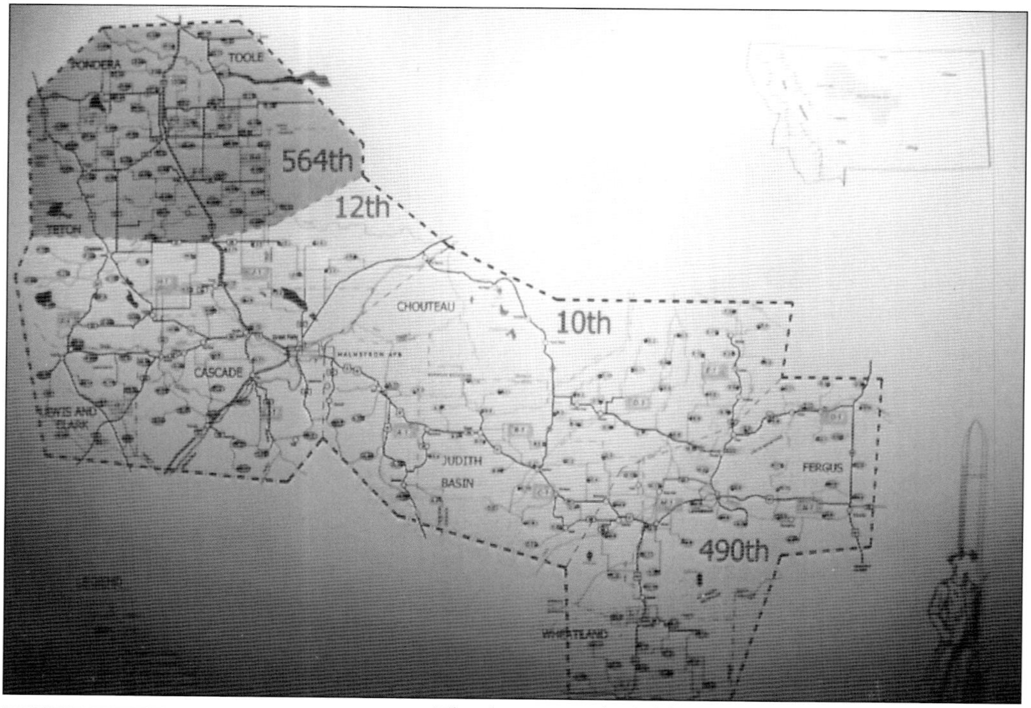

This large missile field map shows the 150 ICBM launch facilities and 15 missile alert facilities (each with an underground launch command center) for the 341st Missile Wing at Malmstrom AFB. It is on display at the Malmstrom Air Force Base Museum. (Author's collection.)

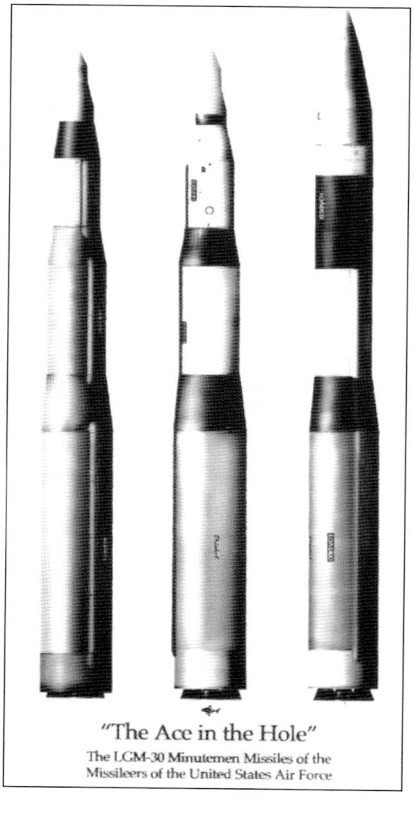

Pictured are the three variants of the Minuteman ICBM. From left to right are I, II, and III. Pres. John Kennedy, after the conclusion of the Cuban Missile Crisis, said the backing down of Soviet premier Nikita Khrushchev from the brink of nuclear war was because of the 10 Minuteman I ICBMs on alert at Malmstrom Air Force Base. This artist's depiction is titled "The Ace in the Hole," a reference to President Kennedy's statement. (Courtesy John Mollison.)

This map shows the locations of the four 341st Missile Wing squadrons. The five flights at upper left, 564th Missile Squadron (1962–2009), are deactivated, and the sites are destroyed. As of 2017, the wing operates three missile squadrons of 50 Minuteman IIIs each. This image provides an excellent view of how large an area the 341st Missile Wing covers in Montana. (Courtesy 341st Missile Wing Public Affairs.)

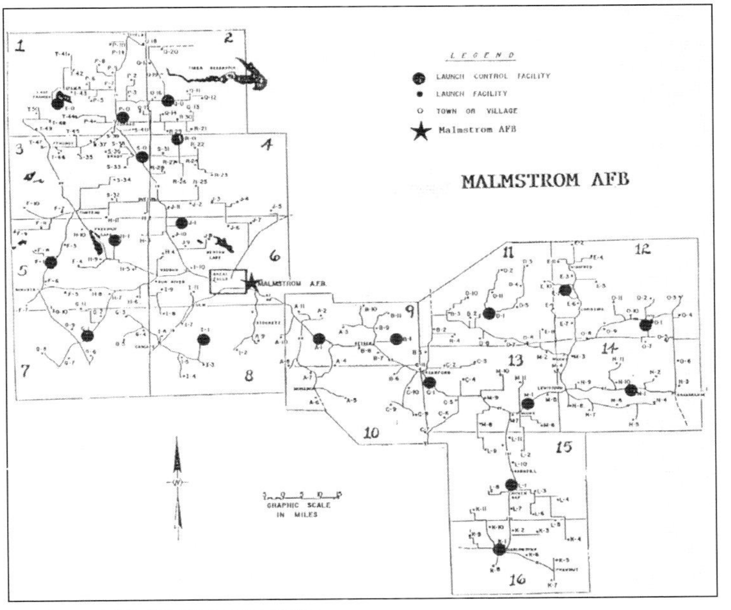

The design of the Minuteman called for a three-stage, solid-fuel missile that was to be extremely reliable, quick to launch, have a high capability for survival and maximum simplicity, and be able to remain on alert in its underground silo around the clock for many years. In comparison, the Atlas F, a liquid-fueled missile, had to be raised out of its silo on a heavy elevator/launch platform. The Titan I had to be launched from the surface, but the Titan II could be fired from its silo because of storable liquid fuel. (Courtesy Bruce Townsley.)

Cables were placed underground for communication between missile facilities. The preferred and fastest system for trenching was by a continuous rotary clamshell excavator that dug out and removed dirt from the trench, digging down to the required depth and piling the dirt to the right of the excavator. Cable construction angered many Montana ranchers and farmers due to destruction of fences and gates, ruts in fields, ruined crops, scattered livestock, and debris from construction crews. (Courtesy 341st Missile Wing.)

After trenching, a truck with a large cable reel slowly drove alongside the trench, with workers carefully unrolling the cable and positioning it into the middle of the trench bottom to prevent any damage. Once this was complete, a bulldozer slowly filled the trench. The disturbed land was returned to the surrounding terrain contours, although many landowners disagreed. (Courtesy 341st Missile Wing.)

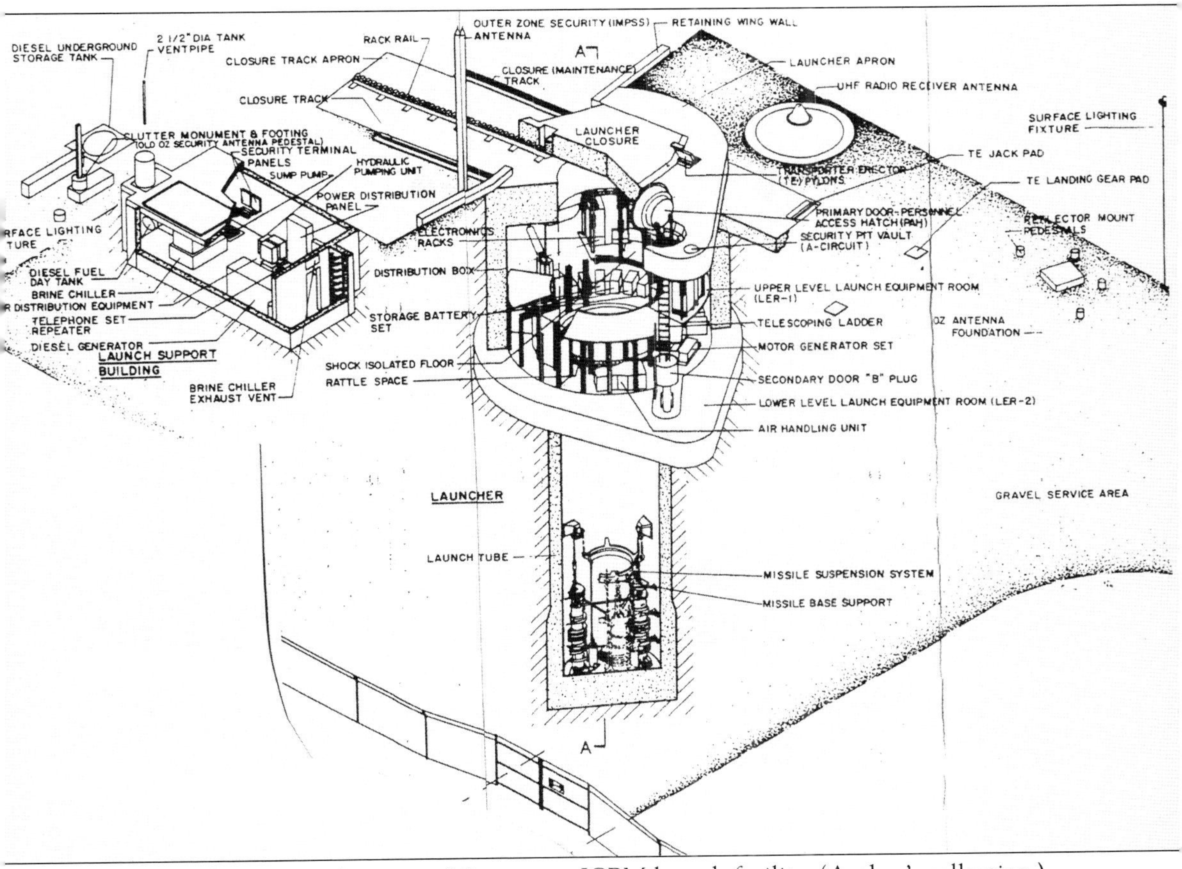

This cutaway diagram is of a remote Minuteman ICBM launch facility. (Author's collection.)

Construction at a launch facility site started with earth scrapers digging a 12-foot excavation, large enough to allow a deeper excavation. Removed earth and rock were piled on the flat ground above and away from the excavation, to be used later for backfill. A crane fitted with a clamshell bucket continued digging to a depth of 32 feet. (Courtesy 341st Missile Wing.)

To allow heavy equipment to access the 32-foot level, an access ramp was excavated. Each silo required a massive amount of excavation work. The slope, width, and depth of the access ramp can be gauged by the two engineers supervising the work. All construction workers were required to wear safety helmets in the work area. (Courtesy 341st Missile Wing.)

Depending on the soil type, a heavy wheeled crane with a large rotary digger, set up over the silo's precise position, began digging a shaft to a depth of 84 feet. The dirt was pulled to the top of the shaft, removed by front end loaders into dump trucks, hauled to the surface, and dumped. This was a slow process requiring constant attention not to over-torque the drilling shaft. (Courtesy 341st Missile Wing.)

Heavy cranes lower a fabricated steel rebar silo form to be installed into the excavated shaft, where it will sit on a prepared concrete base. (Courtesy 341st Missile Wing.)

Reinforced concrete was poured, filling the rebar around the tube, and was then allowed to harden. The tube was backfilled to the 32-foot depth. Workers continually tamped the earth as it was backfilled. The process could not be hurried if the desired hardness of the compacted soil was to be achieved. The silo opening would then be covered with a metal cap, with rebar into the concrete for construction up to the surface level, so as not to damage the launch tube. (Courtesy 341st Missile Wing.)

During construction, a large steel rebar reenforced poured concrete cylinder slowly rose to the surface level. The forms moved upward as the concrete cured, allowing the weight of the added wet concrete above. After the concrete was fully hardened, the excavation was backfilled and tamped. (Courtesy 341st Missile Wing.)

Ground level construction covered the launch tube with a 180,000-pound concrete silo cover, fitted with a pointed and angled end, away from the primary level of the launch facility. It was designed to clear away snow and/or debris from the silo when the door was opened for a missile launch. (Courtesy 341st Missile Wing.)

After construction was complete, interior equipment was installed. (Courtesy 341st Missile Wing.)

An operational Minuteman III launch facility is a non-obtrusive installation, secured with a chain-link fence and security and remote lighting, with 99 percent of the installation underground in a hardened facility. (Author's collection.)

The 341st Missile Wing, 10th Missile Squadron, marked its MAF/LCC at the entrance with a sign marking its historical importance to the Cold War. The Alpha-01 is marked as "the first Ace in the Hole. In continuous operation for over 50 years." This is along North Dakota Highway 3, twenty-four miles southeast of Malmstrom AFB. It was operational on August 2, 1961. (Author's collection.)

Pictured is the Alpha-01 missile alert facility, the first missile facility to go on 24-hour missile deterrent alert, during the Cuban Missile Crisis. It controlled 10 Minuteman III ICBMs. (Author's collection.)

For maintenance inside a launch tube, the silo cover is slowly winched open. This requires a full missile maintenance crew. Note the security forces armed guard in the background protecting the now opened launch tube. Two missile maintenance personnel, each wearing a safety harness to prevent an accidental fall into the open launch tube, finish the canvas environmental cover over the missile below. The thickness of the silo cover can be judged by the airmen in front. (Courtesy 341st Missile Wing Public Affairs.)

The personnel/maintenance access hatch is open to allow missile maintenance personnel into the launch tube interior for access to the missile. (Courtesy 341st Missile Wing Public Affairs.)

In winter, outdoor temperatures can be dangerous when conducting work on the launch facility. (Author's collection.)

After a heavy snow, if maintenance work is required at a remote launch facility, snow removal equipment comes from the associated missile alert facility to clear snowdrifts from the work area inside the launch facility. (Courtesy 341st Missile Wing.)

Pictured is a transporter-erector in full 180-degree upright position over an open launch tube. The warhead and guidance package has been removed. The missile inside the launch tube is then carefully raised into and secured inside the transporter-erector for the drive back to Malmstrom Air Force Base in a protected convoy. (Courtesy 341st Missile Wing.)

Pictured is a test launch of a Minuteman from the 30th Space Wing's intercontinental ballistic missile test and launch range. The missile impacted into a designated target area near Kwajalein after launch. (Courtesy 341st Missile Wing.)

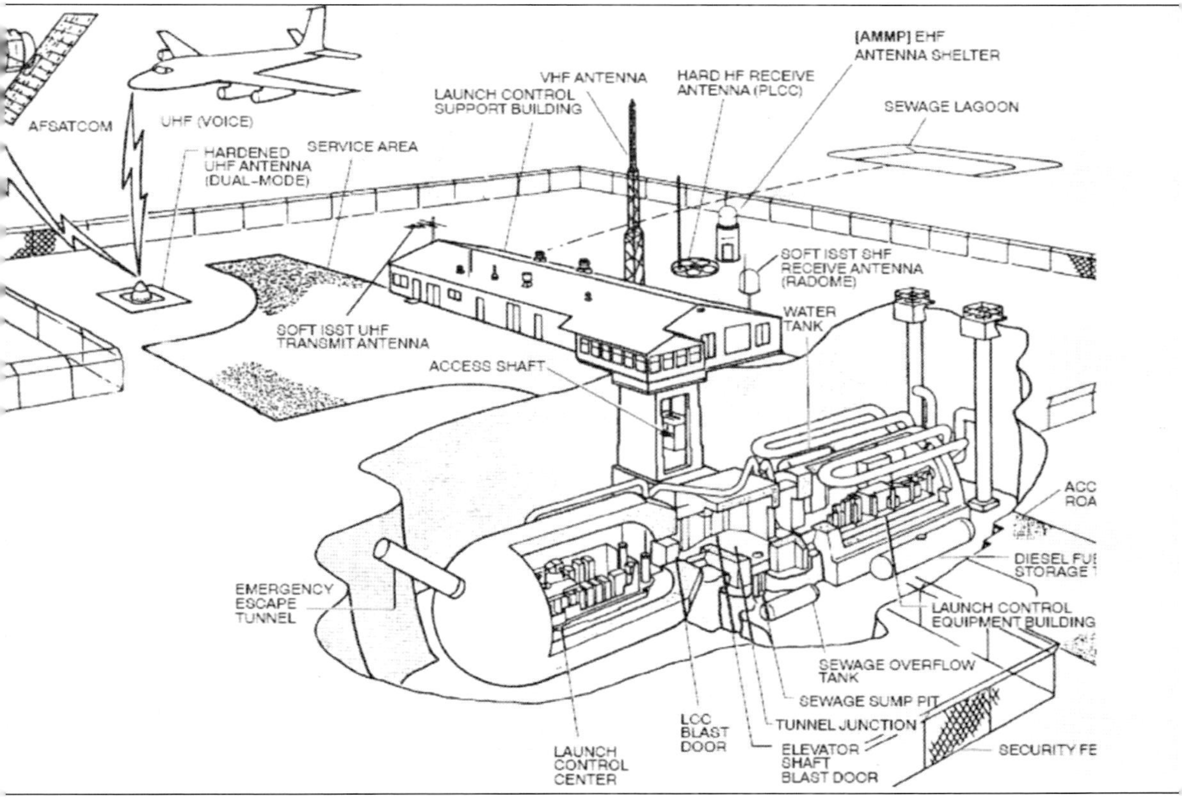

This diagram shows a missile alert facility and its launch control support building, with the underground launch control center and associated launch control equipment building. (Courtesy Minuteman National Historic Site.)

Shown is a launch control support building at the aboveground missile alert facility that supports the underground launch command center. (Courtesy 341st Missile Wing Public Affairs.)

This is an aerial view from a 40th Helicopter Squadron UH-1N as it approaches Lima-01. Communications antennas are visible at center. The sewage lagoon is to the right, outside the security area. The helicopter landed on the concrete pad at left outside the security area. (Courtesy 341st Missile Wing Public Affairs.)

The author interviews security forces personnel at Lima-01 in the missile alert facility general area about their security functions for the site and non-isolated unmanned Minuteman III launch facilities. (Courtesy Great Falls, Montana, Malmstrom Air Force Base, 341st Missile Wing Public Affairs.)

Seen here is the dining area with serving console, microwave, and refrigerator for personnel on duty assignment. (Courtesy 341st Missile Wing Public Affairs.)

This kitchen has been completely modernized and upgraded. An onsite chef prepares individual ordered meals from a menu, rather than heating prepared meals. (Courtesy 341st Missile Wing Public Affairs.)

A chef cooks an ordered meal at Lima-01, which allows onsite personnel a variety of balanced restaurant-type meals while on duty away from Malmstrom AFB. (Courtesy 341st Missile Wing Public Affairs.)

Another upgrade for missile site support crew personnel was an onsite unclassified computer for e-mail correspondence in a separate room, so personnel can stay in touch with family members. (Courtesy 341st Missile Wing Public Affairs.)

Pictured is one of the bunk rooms for onsite personnel, including overnight security forces and maintenance personnel. There are separate bedrooms for male and female personnel, along with separate bathrooms. Part of the upgrades were initiated by the Air Force Global Strike Command. (Courtesy 341st Missile Wing Public Affairs.)

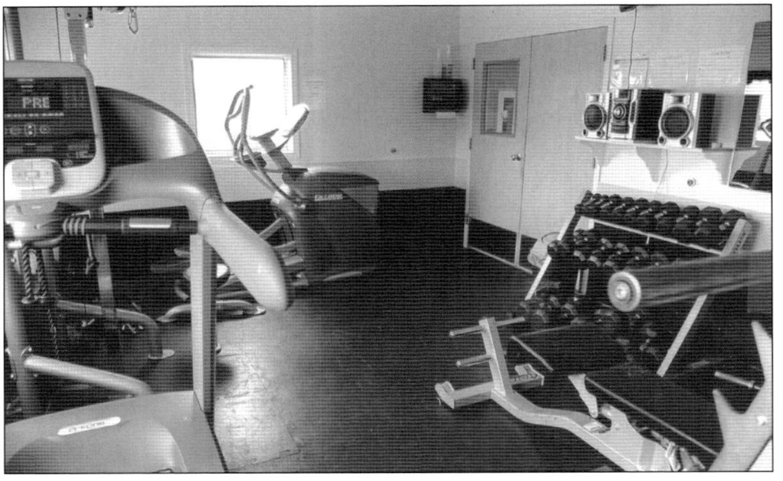

The missile alert facility upgrade included a fully equipped exercise room for onsite personnel to maintain their physical fitness and reduce stress. (Courtesy 341st Missile Wing Public Affairs.)

The construction of the launch command center (foreground) and associated launch control equipment buildings started with excavation to a depth of 31 feet, which was enlarged to provide for construction work. An access ramp was excavated from the surface level to the worksite level. Steel and rebar work are visible in this photograph. The central corridor between the two, which will contain the freight-type elevator for the movement of missile crew and cleared personnel from the surface to the underground complex, will be added. Once these structures were completed, construction work started on the aboveground launch control support building. (Courtesy 341st Missile Wing.)

The freight elevator shaft is pictured under construction, with concrete forms slowly raised as rebar work is completed and concrete is poured and allowed to cure. (Courtesy 341st Missile Wing.)

The launch control equipment building is in the foreground with the launch command center behind it, showing their shells covered with hardened concrete. At lower left is the underground diesel fuel tank for the launch control equipment building's backup electric generators. (Courtesy 341st Missile Wing.)

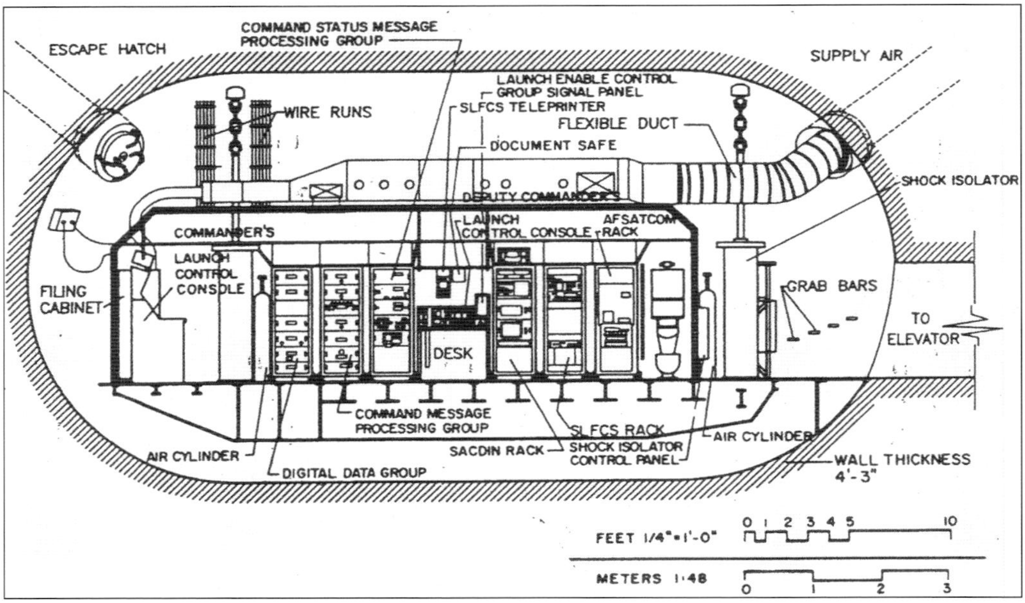

This cutaway diagram shows the interior arrangement of a launch control center. (Courtesy 341st Missile Wing Public Affairs.)

This missile combat crew, with 2nd Lt. Wesley Griffith (left) and 1st Lt. Katie Grimley, was assigned to 10th Missile Squadron, Alpha-01 Launch Command Center, and are pictured on July 15, 2014. This is an excellent view of combat crew seating, new capsule floor carpeting, and electronic equipment in racks. Each crew monitors 10 Minuteman III missiles assigned to the squadron. (Courtesy US Air Force.)

Seen here is an eight-ton launch control center blast door. The door is manually opened and closed, and is perfectly balanced to allow one crew officer to operate it. (Author's collection.)

Seen here is the interior of the launch control equipment building at the Ronald Reagan Minuteman Missile site, a state museum in North Dakota. It has been preserved by the North Dakota State Historical Society after Minuteman alerts ended when the missile wing was inactivated and launch facilities were destroyed to meet Strategic Arms Reduction Treaty limitations. (Author's collection.)

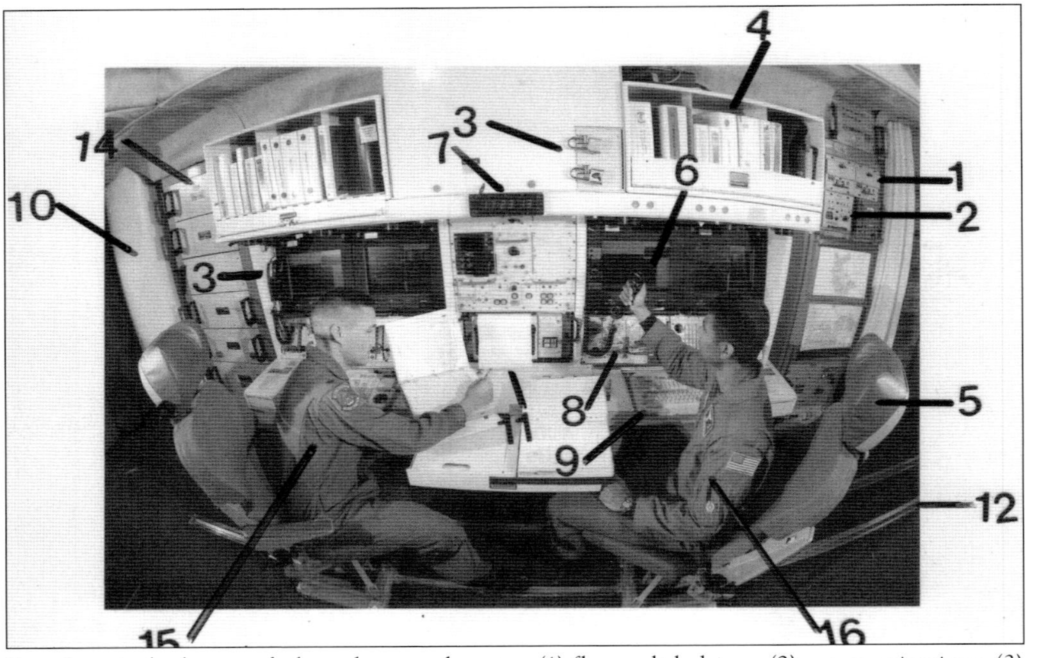

Seen here is the layout of a launch control center: (1) floppy disk drives, (2) communications, (3) launch codes secured by a lockbox with separate missile combat crew officers, (4) manuals, (5) crew chair, (6) APQ computer screen, (7) 24-hour digital clock, (8) missile launch switch, (9) computer keyboard, (10) cover over one of four shock absorbers that suspend the crew compartment within the steel shell, (11) printer, (12) chair transit rails, (13) crew phone, (14) electronic equipment rack, (15) combat crew commander, and (16) deputy combat crew commander. This image was captured with a wall-mounted fisheye lens. (Courtesy 341st Missile Wing Public Affairs.)

The missile combat crew reviews procedures for receiving emergency action messages, authentication, and possible execution. (Courtesy 341st Missile Wing Public Affairs.)

This is the inactivated 564th Missile Squadron launch facility S-31, located 13.4 miles north of Dutton, Montana. An asphalt access road leads to the inactivated launch facility security gate. The site remains physically secured with a padlocked gate and periodic inspection by 341st Missile Wing security forces. However, there are no remote video monitors, alarms, or lights, nor maintenance. (Author's collection.)

This photograph was taken at launch facility S-38 during deactivation. The Minuteman is in the process of being slowly removed from the launch tube. This effort required many specialized vehicles and the ever-present security forces guard. (Courtesy 341st Missile Wing.)

On May 26, 2016, ten two-officer combat missile crews received their pre-departure briefing in wing headquarters before departing for their respective missile alert facility/launch control centers. (Courtesy 341st Missile Wing Public Affairs.)

Missile combat crews travel to their duty missile alert facilities in white, unmarked pickup trucks with matching cargo caps. (Author's collection.)

After checking out their pickups, most missile alert crews stop at the base's exchange to purchase snacks and food for their duty tour. (Author's collection.)

The author is saluted by UH-1N flight engineer Sr. Amn. John Klusek on 40th Helicopter Squadron's aircraft ramp, prior to the flight to Lima-01. (Courtesy 341st Missile Wing Public Affairs.)

This UH-1N flight crew transported the author to Lima-01. From left to right are flight engineer Sr. Amn. John Klusek, pilot Capt. Lucas Richardson, and copilot Capt. Drew Hegarty. (Courtesy 341st Missile Wing Public Affairs.)

Helicopters are the only type of aircraft remaining operational at Malmstrom Air Force Base. This is the view from a UH-1H helicopter after takeoff from Malmstrom AFB en route to missile alert facility Lima-01, northwest of the base. (Courtesy 341st Missile Wing Public Affairs.)

This is a view from inside Lima-01's security area at the closed gate. The two white tanks outside the security gate to the left are fueling tanks. (Courtesy 341st Missile Wing Public Affairs.)

The author receives a briefing on the security forces' daily mission. Security forces personnel maintain site security and conduct remote monitoring of launch facilities, responding to launch facilities to investigate alarms. (Courtesy 341st Missile Wing Public Affairs.)

Seen here is a freight elevator from the security control area to the underground launch control center and launch control equipment building. (Courtesy 341st Missile Wing Public Affairs.)

A 40th Helicopter Squadron UH-1N waits on Lima-01's landing/take-off pad to return to Malmstrom Air Force Base. (Courtesy 341st Missile Wing Public Affairs.)

During an exercise, three tactical response force airmen exit out the open door on the left side of a helicopter with weapons at the ready position, with the fourth exiting out the right door and proceeding around the helicopter's nose to take up a support/flanking position. This action is rapid and provides an overwhelming force to secure a threaten nuclear installation. (Courtesy 341st Missile Wing Public Affairs.)

Adjacent to the open grass landing area is a full-scale surface replica of a launch facility used for training for unauthorized entries. The tactical response force trains on this scenario and is fully capable and equipped to perform this security operation. (Courtesy Great Falls, Montana, Malmstrom Air Force Base, 341st Missile Wing Public Affairs.)

A small intercontinental ballistic missile hard mobile launcher is on display at the National Museum of the US Air Force. (Author's collection.)

A load weight test using a vehicle scale is utilized to accurately calculate per-wheel load weight. This data is used to compare the Montana Highway Department's maximum load weight limits on primary, secondary, and county roads (many of which are gravel surface), over which the missile carrier could drive to designated, pre-surveyed remote launch locations to support nuclear deterrence. (Courtesy 341st Missile Wing.)

Seven

40th Helicopter Squadron 1948–Present

The 40th Helicopter Squadron falls under the 582nd Helicopter Group. 20th Air Force, F.E. Warren AFB in Wyoming. It performs priority security response for the missile field, aerial convoy surveillance, and priority airlift support for missile maintenance and operations. Each successful security operation is the result of the combined efforts of pilots, flight engineers, maintenance crews, and attached security forces personnel providing services under the motto of "Deter, Detect and Defend." Additionally, the 40th supports the National Search and Rescue Plan for Montana and northern Wyoming. Since March 1968, the unit has logged more than 145,000 accident-free hours flying the UN-1N Huey helicopter and is credited with more than 414 saves as of May 13, 2016. The helicopter squadron keeps air activity alive on Malmstrom AFB after the runway was inactivated, leaving only the helicopter landing pad and a drop zone on the runway for C-130Hs from the 120th Airlift Wing at Great Falls International Airport. The primary helicopter pad is at an elevation of 3,472 feet. Helipad No. 1 is a 100-by-100-foot asphalt surface; Helipad No. 2 is the same size but a concrete surface. The 40th Helicopter Squadron, assigned to Malmstrom AFB, began its operational history as Detachment 5 of the 37th Air Rescue and Recovery Squadron and was one of seven detachments under Military Airlift Command. Malmstrom has had helicopters assigned since December 1964 under the Strategic Air Command. The 40th Rescue Flight was activated on May 1, 1993, Air Combat Command, Air Force Space Command, then later transferred to Air Force Global Strike Command. In April 1998, the unit was redesignated as 40th Helicopter Flight, and in October 2005, it was redesignated 40th Helicopter Squadron, ensuring strategic security by providing flexible, rapid-response airlift support to the 341st Missile Wing, aerial surveillance of Department of Defense strategic weapon convoys, and short notice emergency security forces responses and supporting emergency war order tasks and priority personnel and logistical transportation.

The Sikorsky S-55/H-19B helicopter was first used by the Air Force for search and rescue, earning its life-saving reputation. It is pictured on the aircraft ramp at Malmstrom Air Force Base. It was a large helicopter, considered by its pilots to sometimes be underpowered, but served to protect the missile field. It also was slow, at only 110 mph maximum speed; its cruising speed was well below this. (Courtesy 341st Missile Wing.)

During an open house at Malmstrom Air Force Base, a Sikorsky S-55/H-19B helicopter is on display, with its engine compartment open and maintenance steps deployed to allow visitors to look inside the cockpit. (Courtesy 341st Missile Wing.)

UH-1Fs are shown in front of former Air Defense Command alert hangars on Malmstrom Air Force Base. These helicopters equipped the 37th Air Rescue and Recovery Squadron. They had a maximum speed of 148 mph and an operating maximum range of 380 miles. These helicopters served the missile wing in security and transport roles. (Courtesy 341st Missile Wing.)

UH-1Fs with the 37th Air Rescue and Recovery Squadron are pictured on the Malmstrom Air Force Base parking apron. (Courtesy 341st Missile Wing.)

The UH-1Fs operated in all weather conditions, as indicated in this photograph of winter training. The crew and medical personnel are carrying a litter with an injured individual to the helicopter for transport to the nearest medical facility. (Courtesy 341st Missile Wing.)

This close-up view shows the UH-1F in an off-base landing zone with two pilots. The aircraft was highly versatile, able to land in small clearings, even on roads, as needed during rescue operations. (Courtesy 341st Missile Wing.)

This side view shows a UH-1F on its parking ramp, with a KC-135A Stratotanker in the background, when Malmstrom's runway was operational for Strategic Air Command aircraft. (Courtesy 341st Missile Wing.)

Pictured is the left side profile of a UH-1F, the type used on Malmstrom Air Force Base, on the south side of the entrance to the base. (Author's collection.)

Shown is a UH-1F helicopter being mounted on a display post at the Malmstrom Air Force Base Air Park and Museum. (Courtesy 341st Missile Wing.)

This is a view out the right-side cargo/passenger door of a UH-1N approaching the 40th Helicopter Squadron's parking ramp. One UH-1N is tied down on the parking ramp, with the squadron's operations building/headquarters in the background to the right. (Courtesy 341st Missile Wing.)

Pictured is a modern multi-bay helicopter maintenance and storage building. Inside storage is important because of Montana's extreme weather conditions, especially in winter. (Author's collection.)

An alert-ready helicopter is shown on the alert pad. If notified by the command post, the flight crew heads to the aircraft and prepares to take off for a search and rescue mission or transport of a security forces tactical alert force to respond to an unauthorized entry into one of the wing's 150 launch facilities or 15 missile alert facilities. (Author's collection.)

The UH-1N is vital for security escort of Minuteman III missiles and guidance/warhead package movement to and from the wing's 150 launch facilities, carrying a select tactical alert force to support the ground convoy. (Courtesy 341st Missile Wing.)

A UH-IN has just been towed out of its hangar on to an alert pad. The rotor blade is still secured front and back. The aircraft's doors are open. Two maintenance personnel are with the aircraft. (Author's collection.)

A UH-1N is shown on the 40th Helicopter Squadron's parking ramp. The aircraft is being prepared for a flight. (Author's collection.)

One of the requirements for UH-1N crews is to assist in the training of the security forces tactical alert force, usually a four-person team, as seen here. They exit the aircraft, two to each side, with weapons drawn to advance to the launch facility. (Courtesy 341st Missile Wing.)

This is a side view of a UH-1N helicopter with the 40th Helicopter Squadron. The helicopter can carry a maximum of 13 passengers at a maximum gross weight of 10,500 pounds. It has a range of 300 miles with a maximum speed of 145 mph, although it operates at lower cruising speed to reduce wear and tear on the engine and increase operating efficiency, as when flying from Malmstrom AFB to the missile wing's facilities. If an emergency requires, higher speed is used. (Courtesy 341st Missile Wing.)

Pictured is a secure storage area with a UH-1F. (Author's collection.)

Shown is the Sikorsky UH-60G Pave Hawk, an upgraded variant of the US Army Black Hawk. It can carry more personnel than the UH-1N, as well as heavier loads, and uses a sling for oversized cargo. It has a maximum speed of 184 mph. This UH-60 belongs to the South Dakota Army National Guard at Rapid City, South Dakota, based at the Rapid City International Airport. It was suggested as a replacement for the Air Force's UH-1N fleet. This option was not taken. (Author's collection.)

Boeing and Italy's Leonardo won the US Air Force's UH-1N Huey replacement competition and will produce a militarized version of the Italian firm's commercial AW139 helicopter, now called the Gray Wolf. The MH-139 will meet Air Force requirements to carry nine fully equipped troops for an unrefueled range of 225 miles. (Courtesy US Air Force.)

Eight

Red Horse Squadron 1962–Present

The 219th Red Horse Flight, Montana Air National Guard, was established on September 3, 1997, and stationed at Malmstrom Air Force Base. The 219th Red Horse Flight (RHF) was made an associate unit of the 819th Red Horse Squadron, reactivated on August 7, 1997. The 219th RHF provides one third of the manpower and equipment of the 404-person 819th Red Horse Squadron. The 819th Red Horse Squadron is a unit of civil engineers at Malmstrom AFB, responsible for heavy-duty repairs around the world and in the United States, especially at sensitive military facilities such as the 341st Missile Wing, such as new water wells or helicopter landing pads. Activated at Abilene, Texas, it has been inactivated and activated at several different bases over the last 55 years. The unit was most recently activated as the first-ever Air Force National Guard Red Horse associate unit at Malmstrom AFB. The unit is ready to deploy anywhere in the world and remain stationed for an indefinite time. The mission of the 819th Red Horse Squadron is to organize, train, and equip its 120 personnel for its role as a highly mobile, rapidly deployable, self-sustaining heavy construction and repair unit to remote and austere operating environments. The 219th RHF rounds out the 819th Red Horse Squadron with a heavy vertical construction capability, but also maintains proficiency in horizontal construction. Although the 219th RHF is a relative newcomer to the Red Horse mission, it is a long-established Air National Guard civil engineer unit with a distinguished past in service to both national and state responsibilities. The 219th RHF has its roots in the 120th Civil Engineer Squadron assigned to the 120th Fighter Wing (now 120th Airlift Wing) at Great Falls International Airport. Training is the primary peacetime duty of the 219th RHF. Great emphasis is placed upon training in order to maintain the unit in a high state of combat readiness and preparation for operations involving natural disasters and other contingencies.

This red horse statue stands in front of the Donald G. Nutter Hall, headquarters for the 819th Red Horse Squadron. (Author's collection.)

The Red Horse Squadron metal shop is for the repair and fabrication of equipment for deployment operations. (Author's collection.)

More Red Horse Squadron metal shop equipment is seen here. (Author's collection.)

In this concrete testing laboratory, Red Horse Squadron technicians test the strength of concrete that might be used on projects such as runway repair or building foundations. (Author's collection.)

Pictured are squadron deployment bags inside the airfield section, in steel vertical storage racks. The section is inside a converted and modernized former KC-135 maintenance hangar. It allows the pre-positioned personnel equipment to be ready at short notice for rapid deployment. (Author's collection.)

This former Cold War Boeing KC-135 Stratotanker hangar has been modified into a storage facility. The closed hangar door show the padded circle opening, which closed around the tanker's rear fuselage forward of the tail to keep out weather while conducting maintenance on the tanker. (Author's collection.)

This former Cold War Boeing KC-135 Stratotanker hangar is now closed and insulated for the Red Horse Squadron's activities. (Both, author's collection.)

This outdoor vehicle storage and staging area is for the Red Horse Squadron's equipment and containerized supplies. (Author's collection.)

These transportable shelters are used by the Red Horse Squadron as temporary storage or, if anchored on a permanent foundation, as extended storage or for other functions as required. (Author's collection.)

Nine

Weapons Storage Area 1962–Present

The Malmstrom Air Force Base weapons storage areas were designed to safely secure nuclear weapons for deterrence operations, now supporting the 341st Missile Wing's intercontinental ballistic missiles. The complex began with the Strategic Air Command, now part of Air Force Global Strike Command. Physical security begins with a double-perimeter fence enhanced by a security tower, automated intrusion sensors, helicopter anti-landing pole, high-intensity security lighting, and remote video monitoring. The weapons storage area is in a separately secured and isolated area, with access limited only to authorized personnel. This level of security is required to maintain, transport, test, and refurbish nuclear weapons. This is an expensive operation, with constant updates to disassemble and reassemble Minuteman III warheads during modernization and repair. The Air Force mans this facility with highly trained technicians equipped with one-of-a-kind special equipment and manuals in a modern and efficient maintenance area. Convoys to the remote Minuteman launch facilities begin and end at this facility, under heavy ground and helicopter escort to maintain the security of the nuclear weapons. The weapons storage area will eventually be very busy, as Minuteman III ICBMs will be inactivated, removed from their launch facility, moved to the storage site, and replaced by the new generation of ICBM currently under development: the ground based strategic deterrent (GBSD) intercontinental ballistic missile.

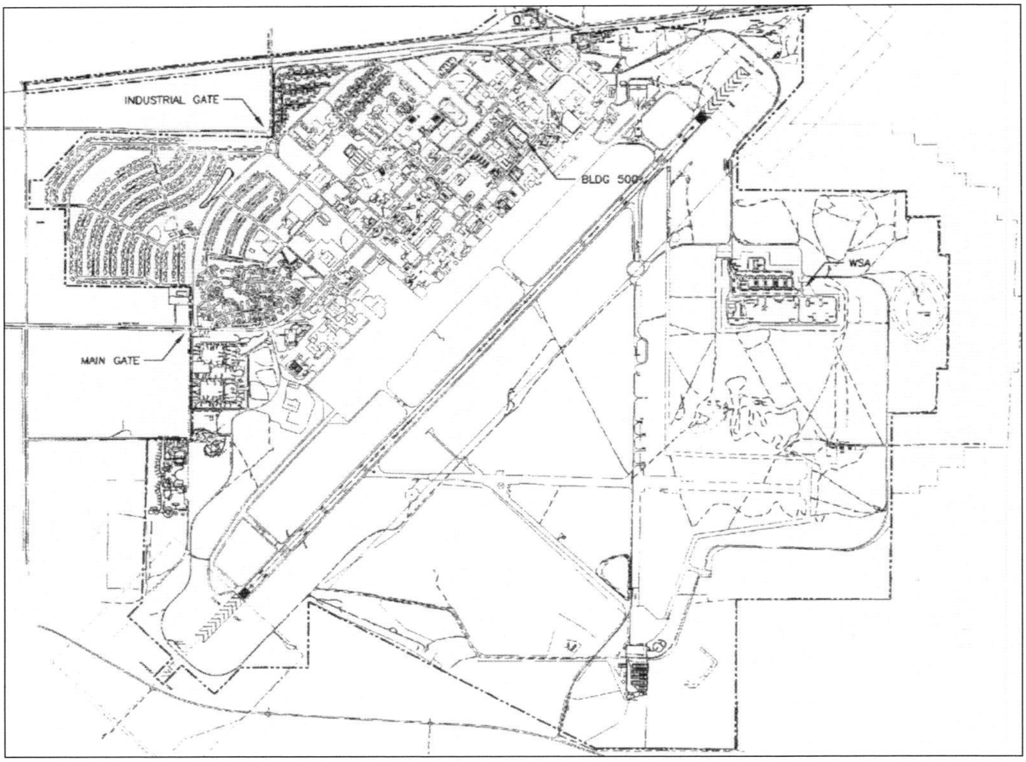

This map shows the isolated location of the weapons storage area to the east of the now non-operational runway at Malmstrom Air Force Base. (Courtesy 341st Missile Wing.)

The security police and tactical response team uses armored vehicles to maintain the security of the weapons storage area and respond to intrusion alarms at the 341st Missile Wing's 150 isolated Minuteman III launch facilities and 15 missile alert facilities. (Author's collection.)

The weapons storage area features double fencing to allow vehicles to drive in through the exterior opened gate, which is then closed prior to the interior gate being opened. (Author's collection.)

Pictured is the new munitions maintenance facility inside the weapons storage area. The building provides updated munitions capabilities to support the Minuteman III ICBM. (Courtesy US Air Force.)

This is a close-up view of the double security fencing at the weapons storage area at Ellsworth Air Force Base, similar to that at Malmstrom. (Author's collection.)

Ten

Overview of Malmstrom Air Force Base 1942–Present

From its beginnings in World War II, Malmstrom Air Force Base has evolved into a modern Air Force installation, responsible for one third of Air Force Global Strike Command's Minuteman III intercontinental ballistic missiles. Malmstrom's physical appearance has undergone many changes since World War II, with major upgrades since 1995. This has included new construction and renovations to base housing, dormitories, and work areas, along with a new commissary and base exchange. Malmstrom was named the best Air Force installation in the continental United States in 1999. The wing has won many Air Force awards for excellence. Located to the east of Great Falls, Montana, it contains a gated community for accompanied housing, managed by a private management company. Base housing consists of two, three, and four-bedroom floor plans, featuring ample storage, master suites, attached garages, fully equipped kitchens, and fenced yards. Those assigned to Malmstrom AFB have access to community playgrounds and picnic areas, a heated swimming pool, and a dog park, as well as modern base facilities. Visitors can also journey back through the base's history from Great Falls Army Air Force Base, transiting to Great Falls and Malmstrom Air Force Base, with the 341st Missile Wing's intercontinental ballistic missiles through the aircraft displays and exhibits of the Malmstrom Air Force Base Museum.

The main vehicle entrance to Malmstrom Air Force Base has enhanced security to protect the nuclear weapons and the Minuteman missile support facilities on the base. (Author's collection.)

This prominent reminder of the Cold War era, a nuclear aircraft crew alert facility, is now used for base support purposes. (Author's collection.)

Pictured are the unaccompanied enlisted barracks on Malmstrom Air Force Base. (Author's collection.)

There is an Army Air Force Exchange Services shoppette and main exchange at Malmstrom Air Force Base. (Author's collection.)

Pictured is the Malmstrom AFB dining facility, a self-service facility for on-base personnel. (Author's collection.)

Pictured is the base commissary, which provides a selection of groceries for active-duty and retired personnel. (Author's collection.)

This is Malmstrom's physical fitness center. (Author's collection.)

The Malmstrom Clinic serves active duty and retired personnel up to age 65. There is a full range of medical care, except emergency services requiring off-base hospital care. (Author's collection.)

Shown are on-base accompanied enlisted and officer housing units, managed by a private company. (Author's collection.)

The Malmstrom Air Park and Museum exhibit building tells the story of the base from World War II to the present. (Author's collection.)

Bibliography

"40th Helicopter Squadron." Great Falls, MT: Malmstrom Air Force Base, 341st Missile Wing Public Affairs.

"A Brief History of the 341st Strategic Missile Wing and Malmstrom Air Force Base, August 14, 1942–December 1973." Great Falls, MT: Malmstrom Air Force Base, 341st Missile Wing Public Affairs, 2015.

"819th Red Horse Squadron." en.wikipedia.org/wiki/819th_RED_HORSE_Squadron.

"Ground Observer Corps." US Air Force publication, January 1, 1952–June 30, 1952. Montgomery, AL: Air Force Historical Research Agency, Information Systems Division. Copy of Roll No. 44844, November 13, 1945.

"History of Malmstrom Air Force Base." Great Falls, MT: Malmstrom Air Force Base, 341st Missile Wing Public Affairs.

"History of Great Falls Army Air Base." Great Falls Chamber of Commerce.

www.techbastard.com.